EARTH, SEA, AND AIR
A Survey of Geophysical Sciences

This book is in the
ADDISON-WESLEY SERIES IN THE EARTH SCIENCES

JOHN WINCHESTER
Consulting Editor

EARTH, SEA, AND AIR
A Survey of Geophysical Sciences

JEROME SPAR
Department of Meteorology and Oceanography
New York University

SECOND EDITION

ADDISON-WESLEY PUBLISHING COMPANY
READING, MASSACHUSETTS · PALO ALTO · LONDON · DON MILLS, ONTARIO

Copyright © 1965
ADDISON-WESLEY PUBLISHING COMPANY, INC.

Printed in the United States of America

ALL RIGHTS RESERVED. THIS BOOK, OR PARTS THERE-
OF, MAY NOT BE REPRODUCED IN ANY FORM WITH-
OUT WRITTEN PERMISSION OF THE PUBLISHER.

Library of Congress Catalog Card No. 64-25804

Second Printing—June 1967

To Fran

Preface to the First Edition

This book has been prepared from lecture notes for a one-semester course in the geophysical sciences. The course is designed for students in a college of liberal arts as part of a broad survey of science. It is assumed that the student has no knowledge of mathematics, physics, or chemistry beyond that acquired in general courses in high school, and that he does not intend to major in any branch of physical science in college. The course is not intended to serve as a foundation for more advanced courses in geophysics. Rather, its purpose is to give the liberal arts student a general semitechnical survey of the physical state of the earth, so that he may understand some of the momentous events that are taking place today in certain branches of science and technology.

Between July 1957 and December 1958 geophysical scientists engaged in one of the greatest programs of international scientific cooperation the world has ever known. This was the International Geophysical Year, the IGY, a program of worldwide exploration of the physical state of the earth and the atmosphere. The most spectacular events of the IGY, and certainly the most significant from both a scientific and political viewpoint, were the launchings of the artificial satellites beginning with Sputnik I on October 4, 1957. However, the IGY was more than just the beginning of the satellite era. It also marked the beginning of a period of the most rapid development of all phases of geophysical science. The IGY experiment has stimulated the most ambitious and imaginative program of geophysical exploration of our planet. Enormous sums of money are being expended by governments to explore the upper atmosphere with rockets and satellites, to survey the oceans, to open up the uncharted wastes of Antarctica, to bore deeply into the interior of the earth, to study the effects of solar explosions on our atmosphere—all in an effort to learn all we can about our planet, the earth. This book is intended to provide a background for the understanding of some of these geophysical activities.

It is hardly necessary in our time to argue the desirability for every intelligent person, no matter what his field of special interest, to have some knowledge of science. No man can consider himself "educated" if he has not learned some of the basic facts and concepts of science. Today science is woven into the fabric of our social and political life. Political decisions of the greatest importance—disarmament, nuclear weapons testing, the exploration of space, to name only a few—are concerned with scientific matters. The gulf that once separated science and politics is being bridged as scientists are called upon increasingly to contribute to political decisions, while social and political leaders are being forced to learn the once esoteric language of science and technology.

Few areas of science are better known superficially to the intelligent layman today than geophysics, the physics of the earth. Geophysics is the study of the physical state of the earth and its atmosphere. It is a vast region of science, embracing many scientific disciplines and divided into many fields of specialization. Geophysics corresponds roughly to the subject called "Earth Science" in the high schools, but is more extensive. It is concerned with man's total natural physical environment—the earth on which he lives, the waters of the earth, the atmosphere surrounding the earth—and with the relationships between the earth and the rest of the solar system.

In this book we divide the subject of geophysics into four broad areas: the earth as a planet (astronomy and geodesy), the lithosphere or solid earth (geology), the hydrosphere or waters of the earth (hydrology and oceanography), and the atmosphere, the gaseous envelope of the earth (meteorology).

We will begin by considering the earth as a planet in relation to the sun, the moon, and the other planets of the solar system. Then we will examine the gross features of the earth, its size, shape, rotation, and gravity. Next we will study the physical structure of the spheroid, both its surface features, which are the principal concern of conventional geology, and the structure of the earth's interior. (The term geophysics was once used to refer *only* to the exploration of the earth's interior. Although many geologists still cling to this restricted definition of the term, the broad definition adopted in this book is rapidly winning universal acceptance. The International Union of Geodesy and Geophysics and the American Geophysical Union, for example, encompass all branches of geophysical science.) Our study of the solid earth will include the geological history of the earth, the evolution of geological features (geomorphology), earthquakes (seismology), and the magnetism of the earth (geomagnetism).

From the solid earth we turn to the waters of the earth and the subject of oceanography, for the oceans cover more than 70 per cent of the earth's surface. In the final section of the book we devote our attention to the atmosphere and the subject of meteorology. Here we will study the physical properties of the gas in which we live, the causes of wind, weather, and climate, and the upper atmosphere. The role of rockets and satellites in the exploration of the upper atmosphere will be reviewed. We will also examine the effects which cosmic rays and anomalous radiations and particles from disturbed regions of the sun have on the upper atmosphere.

Throughout the book an effort is made to stress the interrelation among the various branches of geophysics, and to show how the same physical principles apply to all the spheres of geophysics, whether solid, liquid, or gas.

Included in the book is a set of six laboratory exercises in geophysical science. The laboratory exercises, each of which requires about three laboratory hours, are designed to illustrate the material in the text. In the course for which this book was written, laboratories are attended biweekly by the students. On this schedule the laboratory exercises will be in phase with the lectures and assigned reading.

It should be stressed that this book presents only a superficial survey of the geophysical sciences. The student is urged to read in the references cited for a more thorough treatment of individual subjects.

The assistance of my colleagues, past and present, in the Department of Meteorology and Oceanography at New York University in the development of this book and the course for which it was intended is gratefully acknowledged. I also wish to thank Mrs. Wilhelmina Lewis and Mrs. Sadelle Wladaver for the typing of the manuscript, and Mr. Harold Seitz for assisting in the preparation of the figures.

New York City   J. S.
May 1962

Preface to the Second Edition

The printing of a second edition of *Earth, Sea, and Air* has provided the author with the opportunity to remove a number of errors in figures and in text, to add some clarifying material, and to make consistent use of the metric system.

I am grateful to the many colleagues, critics, and correspondents who have made this revision not only possible but necessary.

New York City   J. S.
October 1964

Contents

CHAPTER I. THE EARTH AS A PLANET
1. The origin of the earth 1
2. The solar system 5
3. Time and the rotation of the earth 15
4. The shape and size of the earth 21
5. Gravity 28
6. Maps 35

CHAPTER II. THE LITHOSPHERE
1. Geological history 42
2. The earth's crust 49
3. The interior of the earth 54
4. Earthquakes 58
5. Geomagnetism 61

CHAPTER III. THE HYDROSPHERE
1. The hydrologic cycle 68
2. The ocean basins 71
3. The oceans 77
4. Currents, waves, and tides 84

CHAPTER IV. THE ATMOSPHERE
1. Air 99
2. Vertical structure of the atmosphere 102
3. Wind 108
4. Weather 117
5. Solar radiation and climate 123
6. The upper atmosphere 131

LABORATORY EXERCISES
I. The sundial 138
II. Gravity 142
III. Geomagnetism 142
IV. Ocean currents 145
V. Water waves 145
VI. Weather observing 150

INDEX 153

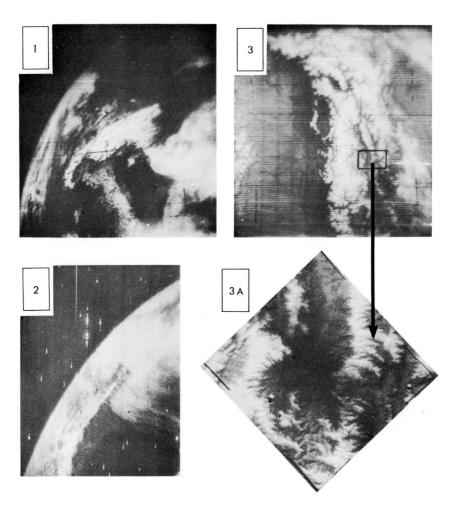

Mountain areas appear prominently in TIROS I photographs. (1) Snow and a few clouds over the Alps (April 2, 1960). Italy extends southward from the Alps to the lower center of the picture. (2) Clouds streaming off South America (April 29, 1960). (3) The snowy Himalayas capped by a few clouds (May 13, 1960). (3a) A high-resolution camera view of the area outlined in (3). (Photographs courtesy U.S. Department of Commerce, Weather Bureau.)

I. The Earth as a Planet

Chapter I is concerned with the planet Earth as a whole, and not with any of its specific features. We begin with a short review of the theories of the origin of the earth and the solar system. Next we examine the place of the earth in the solar system, the relation of the earth to the sun and the moon, the rotation of the earth and its use as a clock, the over-all dimensions of the earth, and the measurement of gravity and its variations. The construction of maps, which are so necessary in geophysics, is the subject of the final part of this chapter.

1. The origin of the earth. It is now generally agreed that the earth was born about 4.5 billion years ago. There is, however, no general agreement as to how the earth (or the sun, the planets, and the moon) formed. We begin this section by reviewing the evidence for the age of the earth, and then go on to examine the theories of formation that are now most widely accepted.

The age of the earth. The determination of the age of the earth was made possible by the discovery of radioactivity in 1896, and the subsequent discovery that the earth contains abundant quantities of radioactive elements. As these radioactive substances "decay," they emit alpha, beta, and gamma rays. (Alpha rays, or particles, are the nuclei of helium atoms; beta rays, or particles, are high-speed electrons; gamma rays are high-frequency radiations similar to x-rays.) The absorption of these radiations within the earth transforms the energy of the radiation into heat and accounts for the high temperatures found below the surface of the earth.

By the emission of alpha, beta, and gamma rays, a radioactive element is spontaneously transformed into another element. This transformation occurs in the atomic nucleus and depends upon the nuclear properties of the particular radioactive *isotope* (or nuclide) of the element.

The isotopes of any given element are atoms which have almost identical chemical properties and occupy the same place in the periodic table of chemical elements. (The word "isotope" is derived from the Greek *iso,* equal, plus *topos,* place.) However, the isotopes of the element differ slightly from one another in their masses. Each of the chemical elements is characterized by the number of protons in its nucleus. This number, which identifies the element in the periodic table, is called the *atomic number.* In addition to the positively charged protons in the nucleus and the equal number of negatively charged electrons outside the nucleus of

TABLE 1
PRINCIPAL RADIOACTIVE ISOTOPES IN THE EARTH

Parent	Stable daughter	Half-life, billions of years
Uranium-238 (U^{238})	Lead-206 (Pb^{206})	4.5
Uranium-235 (U^{235})	Lead-207 (Pb^{207})	0.7
Thorium-232 (Th^{232})	Lead-208 (Pb^{208})	15.0
Rubidium-87 (Rb^{87})	Strontium-87 (Sr^{87})	60.0
Potassium-40 (K^{40})	Argon-40 (A^{40})	1.3

the atom, the atomic nucleus also contains neutral particles, called *neutrons.* The total number of protons plus neutrons in the nucleus of the atom is called the *mass number.* The different isotopes of any element have the same atomic number, but are characterized by different mass numbers. Uranium-238, for example, is an isotope of uranium (atomic number 92) with mass number 238. Uranium-235, whose atomic number also is 92, is a lighter isotope of uranium, with mass number 235.

The decay of a radioactive isotope (or radionuclide) results in its transformation into another element, which, if radioactive, will also decay, etc. This radioactive chain ultimately produces a stable, nonradioactive isotope which does not decay any further. The original radioactive isotope is known as the *parent,* and the isotopes in the radioactive decay chain are called *daughters* or *daughter products.*

Each radioactive isotope decays at a constant and characteristic rate, which is expressed in terms of its *half-life,* the time required for one-half the mass of a radioactive isotope to disappear by decay, i.e., the time required for one-half of the atoms of the radioactive isotope to be transformed into another isotope. The half-life of the radioactive material found in the earth is the clock used to measure the age of the earth. The process of determining the age of rocks, organic material, and archeological remains by measuring their radioactivity is known as radioactive dating.

Table 1 lists the principal radioactive isotopes in the earth, their stable daughter products, and the corresponding half-lives. The age of a rock can be established by measuring the relative quantities of the parent and daughter isotopes found in it. The result is expressed in terms of the isotope ratio, i.e., the ratio of the quantity of daughter product in the rock to the quantity of parent product still left. This ratio increases with the age of the rock. From the isotope ratio and a knowledge of the half-life, the age of the rock can be determined. For example, if the ratio of lead-207 to uranium-235 in a rock is found to be 1, the age of the rock is 0.7 billion years, for it takes 0.7 billion years, the half-life of uranium-235,

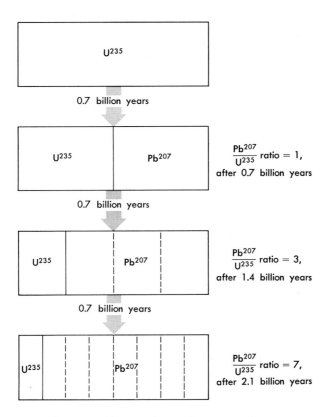

Fig. 1. Radioactive decay of uranium to lead.

for one-half the original amount of uranium to be transformed into lead. At the end of one half-life there are equal amounts of uranium and lead in the rock, and the Pb^{207}/U^{235} ratio is 1. At the end of two half-lives, half the remaining uranium has decayed to lead, and the Pb^{207}/U^{235} ratio has increased to 3; after three half-lives the ratio is 7; etc. (See Fig. 1.)

The oldest measured rocks on earth are about 3.4 billion years old. This, however, is only the time elapsed since these rocks last solidified, and the earth is older than that. From an analysis of the relative abundance of various isotopes of lead in the earth and in meteorites it has been determined that the earth formed into a state similar to what it is today about 4.5 billion years ago.

The origin of the earth. Two theories of the formation of the solar system have competed for many years. One of them, usually referred to as the fragmentation theory, was first proposed by Buffon in 1749. In its various versions, the fragmentation theory proposed that the planets were torn out of the sun when it either collided with another star, or came close enough for the force of gravitational attraction to tear it apart.

The alternative hypothesis, known as the condensation or compaction theory, was first proposed by Kant in 1755, and was worked out in detail by Laplace in 1796. According to this theory, all parts of the solar system formed at about the same time by compaction of a cloud of cosmic dust and gas. Although both theories present difficulties, the condensation theory is presently favored by most scientists, principally because measurements of radioactivity supply further evidence that all parts of the solar system, including the sun, formed almost simultaneously.

There are, similarly, two theories for the formation of the moon. One holds that it was torn out of the earth, possibly from the Pacific Ocean. However, most scientists reject the fragmentation theory of the moon, at least that part which claims that the moon came out of the Pacific Ocean. The other theory argues that the moon formed when the earth formed, and by a similar condensation process.

The modern version of the condensation theory may be described, in simplified form, as follows. A diffuse nebula composed of dust and gas was compressed by the pressure of the light from neighboring stars. The compaction of the nebula caused an increase in the gravitational attraction toward the center of the nebula, which in turn produced a further compression. The increased pressure caused the temperature to rise in the interior of the mass, leading to a thermonuclear reaction and the formation of the sun. The thermonuclear reaction, a transmutation of hydrogen into helium accompanied by a transformation of mass into energy, is responsible for the high temperature and luminescence of the sun.

The cosmic dust left over from the formation of the sun and revolving about it condensed in patches, thus forming the beginnings of the planets. Originally these first planets, called protoplanets (Gr. *protos,* first), were cold. Subsequently they lost their gases and were heated by compression and radioactivity until they became molten. Finally, about 4.5 billion years ago, they solidified.

The earth formed from protoearth by compression, like the other planets, and its moon formed at the same time. As the earth condensed, it heated and melted, the heating being caused partly by radioactivity. When the earth solidified, its core remained in a molten state. The first atmosphere of the earth, composed of gases no longer found in abundance on earth, disappeared. Later the present atmosphere was formed, first by emanations of gases from the earth, and later, after life developed, by exhalations of oxygen from plants. All the oxygen in our atmosphere today has come from plants.

What will happen to the earth? It has been estimated that as the radioactivity in the earth decreases, volcanic activity and mountain building will diminish. Erosion will smooth out the surface, and in about 10 billion years the oceans will cover the earth. But at the same time the sun is expanding and becoming brighter, and in about two billion years it may have grown so much as to be in the orbit of Mercury. The heat from the sun will make the oceans boil. Then the sun will begin to cool and shrink as it uses up its supply of nuclear fuel, and the steam on the earth will freeze. Long before that, life will have ended on the planet.*

2. The solar system. The subject of this section is the relationship of the earth to the sun, the moon, the planets, and the stars. All these astronomical bodies play a role in some branch of geophysics. The sun, as the source of energy and light for the earth, and our reference object for the keeping of time, is of the greatest importance. The gravitational attraction of the moon, which produces the tides in the oceans, causes that object to be of more than aesthetic interest to geophysicists. Although the planets and stars exert a far smaller direct influence on the earth than do the sun and the moon, they are nevertheless of geophysical importance. These astronomical reference points are employed in geodesy, the science of earth measurement, for the precise determination of position, including the practical art of navigation, and in the measurement of the earth's motion.

Historical development. For more than 2000 years the "obvious fact" that the stars, the planets, and the sun, as well as the moon, all move around the earth was accepted as the basis for a *geocentric* (earth-centered) model of the solar system (see Fig. 2a). The apparent motions of the planets, moon, and sun relative to the stars and to one another were explained almost completely in the geocentric theory of Hipparchus (*ca.* 140 B.C.), greatest astronomer of ancient Greece. The theory was further developed by Claudius Ptolemaeus (Ptolemy) about 150 A.D., and is usually referred to as the *Ptolemaic theory.*

In the Ptolemaic theory the earth stood still in the center of the universe. The moon revolved about the earth in the orbit closest to it, while the stars, fixed in a great celestial sphere, revolved in orbits farthest from the earth. Between the moon and the stars lay the orbit of the sun. The planets (Gr. for wanderers), whose motion relative to the stars is reflected in their ancient name, moved in separate orbits about the earth. The orbits of Venus and Mercury were placed between those of the moon and sun, while the orbits of the then-known remaining planets—Mars, Jupiter, and Saturn—were thought to lie between the sun and the stars. The principal difficulty of the geocentric model was the periodic retrogression of the planets. Their apparent paths during the year relative to

---

* A list of references is provided at the end of each chapter. The references are numbered, and a footnote at the end of each section refers the reader to texts of interest. For Section 1, see references 1 through 9 of Chapter 1.

THE EARTH AS A PLANET

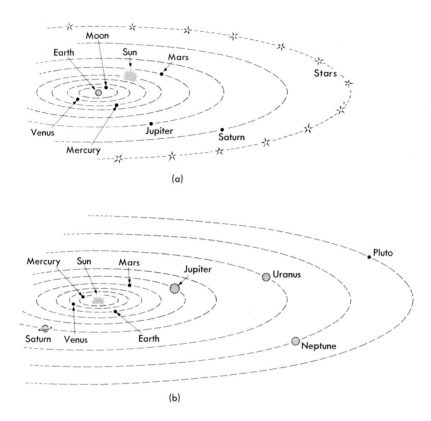

Fig. 2. (a) The geocentric universe. (b) The heliocentric solar system.

the stars were not smooth curves. Instead, they were observed to loop backward occasionally before moving forward again during the year. To explain this apparent backward motion within the framework of geocentric theory, it was necessary to assume that the planets moved in small circular paths, called *epicycles,* as they followed their larger orbits around the earth.

Even among the ancients the geocentric theory was occasionally questioned. Hicetas (*ca.* 450 B.C.) and Herakleides (*ca.* 350 B.C.) taught that the earth rotated on its axis, although they apparently did not question the central position of the earth in the universe. The Greek astronomer Aristarchus (*ca.* 310–230 B.C.), however, did suggest that the sun may be at the center of the universe and that the earth may be revolving about it, although he later rejected the idea. The concept of a *heliocentric* (sun-centered) universe did not gain a permanent place in astronomy until the publication in 1543 of *De Revolutionibus Orbium Coelestium* by Niklas Koppernigk (Copernicus, 1473–1543). Copernicus replaced the complicated geocentric theory, with its planetary epicycles, by the simpler heliocentric (Gr. *helios,* the sun) model in which the sun is at the center of the universe (see Fig. 2b). In the Copernican model the stars, fixed in a celestial sphere, revolve about the sun. Between the stars and the sun are the planets, including earth, revolving about the sun in separate circular orbits. The apparent retrograde motion of the planets is explained by the relative motions of the earth and the planets, which move with different angular velocities about the sun.

The Copernican model was deficient in two respects: the stars do not revolve about the sun, and the planets do not follow circular orbits. The correct description of planetary orbits was given by Johannes Kepler (1571–1630), assistant and successor to the astronomer Tycho Brahe (1546–1601). From an analysis of Brahe's pretelescopic astronomical data, Kepler determined the character of the planetary orbits, and in 1609 he described them in three immortal empirical laws, as follows.

1. The planets move in elliptical orbits about the sun, with the sun at one focus of the ellipse.

2. A line drawn from the sun to any planet sweeps out equal areas in equal times.

3. The ratio of the cube of the average distance of the planet from the sun to the square of its period of revolution about the sun is the same for all the planets.

The invention of the telescope in 1610 and especially the work of Galileo (1564–1642) accelerated the development of astronomy and firmly established the heliocentric model of the solar system. From Kepler's laws, Isaac Newton (1642–1727) was led to the law of universal gravitation (1687), which states that the gravitational force between two bodies is proportional to the product of their masses and inversely proportional to the square of the distance between them. Newton's work showed that Kepler's laws, although originally empirical, conformed to the fundamental laws of motion.

The planets and their orbits. As shown by Kepler and Newton, the orbits of the planets are ellipses. An ellipse may be drawn by placing two thumbtacks in a piece of paper some distance apart and looping a continuous string about them. If a pencil is then placed firmly against the string to form a triangle with the two tacks, the curve drawn by moving the pencil with the string held taut will be an ellipse (Fig. 3a). The two thumbtacks represent the foci of the ellipse. The sum of the distances from the foci to the curve of the ellipse is constant. As the foci come closer together, the ellipse approaches a circle. The departure of an ellipse from a circle is measured by its *eccentricity,* which is the ratio of the distance between the foci to the major diameter of the ellipse (Fig.

TABLE 2
SUMMARY OF SOLAR SYSTEM DATA

	Sun	Mercury	Venus	Earth	Mars	Jupiter	Saturn	Uranus	Neptune	Pluto	Moon
Mean distance from sun, A. U.*	—	0.39	0.72	1.0	1.5	5.2	9.5	19.2	30.1	39.5	384×10^3 km†
Inclination of orbit to ecliptic, degrees	—	7.0	3.4	0	1.85	1.3	2.5	0.8	1.8	17.1	5.1
Eccentricity of orbit	—	0.206	0.007	0.017	0.093	0.048	0.056	0.047	0.009	0.249	0.055
Diameter (earth = 1)	109.1	0.38	0.91	1.00	0.52	10.97	9.03	3.73	3.38	0.45	0.27
Mass (earth = 1)	332×10^3	0.05	0.81	1.00	0.11	318	95.2	14.6	17.3	0.01	0.012
Mean density (water = 1)	1.41	5.46	5.06	5.52	4.12	1.35	0.71	1.56	2.47	2	3.33
Surface gravity (earth = 1)	—	0.30	0.90	1.00	0.38	2.65	1.14	0.96	1.10		0.17
Sidereal period of revolution (sidereal year)	—	88^d	225^d	365.3^d	687^d	11.86^y	29.46^y	84^y	164^y	248^y	27.3^d‡
Period of axial rotation (sidereal day)	24.7^d	88^d	uncertain	23^h56^m	24^h37^m	9^h50^m	10^h02^m	10^h08^m	15^h48^m	6^d10^h	27.3^d‡
Number of natural satellites		0	0	1	2	12	9	5	2	0	

* 1 A. U. = 1 astronomical unit is the mean distance of the earth to the sun, which is ~ 149,504,000 km (92,897,000 miles).
† Mean distance of the moon from the earth.
‡ The symbols y, d, h, m, s denote year, day, hour, minute, and second, respectively.

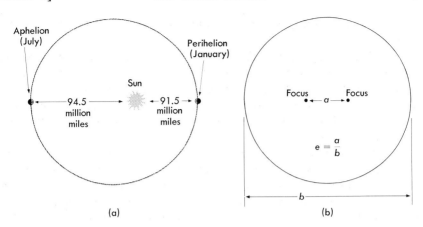

Fig. 3. (a) The orbit of the earth. (b) Eccentricity (e) defined.

3b). The eccentricity of a circle is zero. The eccentricity of the earth's orbit, which is almost circular, is only 0.017.

During the *planetary year,* the time required for the planet to revolve once about the sun, the distance between the planet and the sun changes. When a planet is closest to the sun it is said to be at perihelion (Gr. *peri,* around or next to, plus *helios,* sun), and when it is farthest from the sun it is said to be at aphelion (Gr. *ap,* away, plus *helios*). The earth is at aphelion in July and at perihelion in January. The aphelion distance of the earth is 152 million kilometers (94.5 million miles), and the perihelion distance is 147 million kilometers (91.5 million miles). The mean (average) distance from the earth to the sun is 150 million kilometers (93 million miles). This distance is called an astronomical unit. In Table 2, which lists the planets and some of their characteristics, the distances of the planets from the sun are expressed in astronomical units.

The four planets closest to the sun—Mercury, Venus, Earth, and Mars—are known as the terrestrial planets, and the remaining planets—Jupiter, Saturn, Uranus, Neptune, and Pluto*—are called the major planets. (The planets between the Earth and Sun—Mercury and Venus—are called the inner planets, while those outside the Earth's orbit are known as the outer planets.) Between the orbits of Mars and Jupiter lies the belt of the asteroids, thousands of minor planets and fragments whose origin is still a matter of controversy. All the planets revolve about the sun in the same direction, which is also the direction in which the moon revolves about the earth and the earth rotates about its axis. All the planets except Uranus also rotate in the same direction about their axes. Furthermore, all the planetary orbits except those of Mercury and Pluto lie in almost

---

* It is not certain that Pluto is actually a planet. Some astronomers believe it to be an escaped satellite of Neptune.

the same plane. The plane of the earth's orbit is called the plane of the ecliptic.

It was shown by Newton that the elliptical form of planetary orbits is due to the gravitational attraction between the sun and the planets. Newton proved that the path of any revolving body (e.g., a planet) acted upon by a central force (the gravitational attraction between sun and planet) whose magnitude varies inversely as the square of the distance must be an ellipse.

For the planets to "sweep out equal areas in equal times," as stated by Kepler's second law, it is necessary that they move faster in their orbits when close to the sun (e.g., at perihelion) than when they are far from the sun (e.g., at aphelion). This behavior causes the length of the day to vary slightly through the year. The law of "equal areas in equal time" is a consequence of the fact that the planets conserve their *angular momentum* as they revolve about the sun. Momentum is the product of the mass of an object multiplied by its velocity. Angular momentum is the momentum multiplied by the radial distance of the object from the point about which it is rotating. If a planet conserves its angular momentum about the sun, it must move faster when it is close to the sun than when it is far from the sun. The planets revolve about the sun with unchanging angular momentum because there are no forces in the direction of their motion to cause them to revolve faster or slower. (The gravitational attraction of the sun exerts no net force along the planetary orbit over a planetary year.) Once it has started its revolving motion about the sun, the planet continues to revolve with constant angular momentum unless it is acted upon by some force directed along its orbit. (An artificial satellite outside the earth's atmosphere moves in an elliptical orbit about the earth with constant angular momentum. Within the earth's atmosphere, its angular momentum decreases with time because of the frictional force of the atmosphere.)

Kepler's third law regarding the relation between the distance of a planet from the sun and its period of revolution can be verified from the data in Table 2. We note that the period of revolution, or planetary year, of the planets increases from 88 days for the closest planet, Mercury, to 248 years for the most distant planet, Pluto. To understand Kepler's third law, it is convenient to introduce the idea of centrifugal force. A rotating body behaves as if acted upon by two forces: an inward-directed force, which for a planet is the gravitational force between sun and planet, and an outward-directed force called the centrifugal force. For a body moving in a circle with constant angular velocity, the two forces are balanced. (The angular velocity is the angle swept out by the radius per unit time.) The average angular velocity of a planet is equal to one revolution (360 degrees) divided by its period of revolution, and is virtually constant for each planet. The centrifugal force is proportional to the distance of the planet from the sun multiplied by the square of its angular velocity. The gravitational attraction of the sun is inversely proportional to the square of the planets' distance from the sun. If the gravitational attraction is exactly balanced by the centrifugal force, it can be shown with a little algebra that Kepler's third law must follow. (Optional exercise: Derive Kepler's third law from the balance of gravitational and centrifugal forces.)

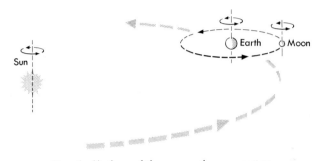

Fig. 4. Motions of the sun-earth-moon system.

The sun-earth-moon system. The revolution of the earth about the sun, and of the moon about the earth, as well as the rotation of the three bodies about their individual axes, take place in the same direction. Looking down on the earth and the solar system from a point above the North Pole, we would observe that the motions occurred in a *counterclockwise* sense, as indicated in Fig. 4.

The largest body in the solar system is the sun, with a diameter 109 times that of the earth, and a mass 333,000 times the earth's mass. The diameter of the moon, 3476 kilometers (2159 miles), is only a little more than one-fourth as large as that of the earth, while its mass is equal to only a little more than one percent of the mass of the earth.

The earth completes one revolution about the sun in space, i.e., relative to the stars, in a period of time called the sidereal* year, which is a few minutes longer than the calendar year.† The moon completes one revolution about the earth in space in a period of time called the sidereal month, which is equal to about 27.3 days ($27^d7^h43^m11.5^s$). This is not the same as the calendar month, nor is it the same as the month based on the phases of the moon, which is discussed below. The sun rotates about its axis, completing one rotation relative to the stars every 24.7 days. The earth completes one rotation about its axis relative to the stars every 23

---

* Sidereal time is "star time," i.e. time measured relative to the stars. The system of time in common use is solar time, which is time measured relative to the sun. Solar time is discussed in Section 3.

† The calendar year will be discussed in Section 3.

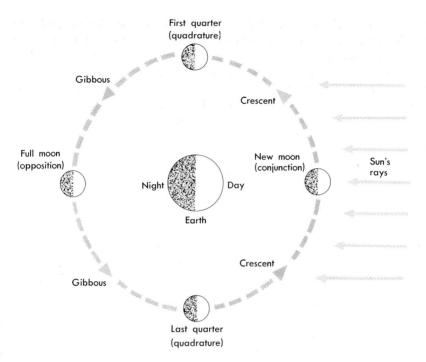

Fig. 5. Phases of the moon.

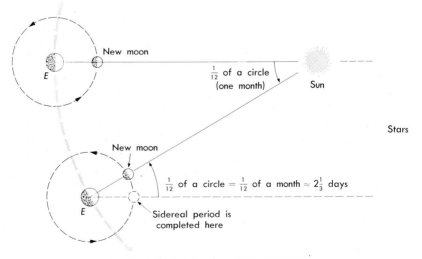

Fig. 6. Sidereal month and synodic month.

hours 56 minutes and 4.09 seconds. This last time interval is called a sidereal day.* Finally, the moon also rotates about its axis, relative to the stars, completing one rotation in exactly the time it takes for the moon to go once about the earth, i.e., 27.3 days. Thus the same side of the moon always faces the earth.

The orbit of the moon is an ellipse with the earth at one focus. At apogee (Gr. *ap,* away, plus *ge,* earth), the point in the lunar orbit farthest from the earth, the moon is 406,000 kilometers (253,000 miles) away, while at perigee (Gr. *peri,* around or near, plus *ge,* earth), its closest approach to earth, the moon is only 356,000 kilometers (222,000 miles) from the earth. The average distance of the moon from the earth is 387,000 kilometers (239,000 miles).

The plane of the moon's orbit is inclined at an angle of 5.1 degrees to the plane of the ecliptic. As a result, the moon does not normally interfere with the light coming to the earth from the sun, nor does the earth usually block the sunlight coming to the moon. The phase of the moon depends on its position relative to the earth and the sun, as shown in Fig. 5. When the moon is between the earth and the sun, it is said to be in conjunction. At this time, only the side of the moon away from the earth is illuminated, and the phase of the moon is "new." The moon then appears above the horizon only during the day, and is dark. After new moon, the moon passes into the crescent phase, the illuminated crescent growing larger each day. When the moon reaches the point of its orbit which is on a line perpendicular to that joining the earth and sun, the moon is said to be in quadrature, and its phase is the first quarter (half moon). After the first quarter the moon passes into the gibbous ("humpbacked") phase, during which it is more than half full. When the moon is on the side of the earth opposite to that facing the sun, it is said to be in opposition. The entire lunar hemisphere facing the earth is now illuminated, and the moon appears full. The full moon appears only on the night side of the earth. Next, the moon passes through the second gibbous stage, through the last quarter, through the crescent stage, and back again to new moon.

The time interval required for the moon to pass completely through all its phases (from new moon to the next new moon, for example) is called the synodic month (from the Latin word for conjunction) and is equal to about 29.5 days ($29^d12^h44^m2.8^s$). This is the interval on which the calendar months are based. The difference between the sidereal month, the lunar period of revolution relative to the stars, and the synodic month, the lunar period of revolution relative to the sun, is illustrated in Fig. 6. The stars may be considered to be infinitely far away. Thus the light

---

* The difference between the sidereal day and the 24-hour solar day will be discussed in Section 3.

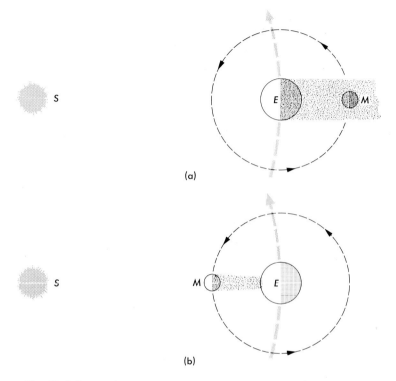

Fig. 7. Eclipses of moon and sun. (a) Lunar eclipse. (b) Solar eclipse.

from the stars may be thought of as entering the solar system along parallel rays. Because the earth is revolving about the sun, the moon must travel farther than one sidereal revolution to complete its synodic revolution relative to the sun. The extra distance corresponds to about one-twelfth of a circle (one month or one-twelfth of a year) in the earth's orbit, and therefore to one-twelfth of a circle or $2\frac{1}{3}$ days in the moon's orbit. Hence the synodic month (from phase to phase) is about $2\frac{1}{3}$ days longer than the sidereal month.

When the moon in its orbit crosses the plane of the ecliptic along a line connecting the earth and the sun, an eclipse occurs. On these occasions the sun, earth, and moon lie along the same straight line. When the earth lies between the moon and the sun, at a time of full moon, the shadow of the earth is cast on the moon, and a lunar eclipse occurs. If the moon lies between the earth and the sun, at a time of new moon, the shadow of the moon is cast on the earth, and a solar eclipse occurs (see Fig. 7). In the present century, about 375 eclipses will have occurred, of which 147 will be lunar eclipses and 228 will be solar eclipses.*

---

* See references 9 through 12.

3. Time and the rotation of the earth. The rotation of the earth about its axis makes the stars appear to move across the sky. This apparent motion of the stars is the basis for sidereal (star) time, which is of interest principally to astronomers. The apparent motion of the sun, also resulting from the earth's rotation, is the basis for solar time, the universal and common system of timekeeping employed in clocks. The rotation of the earth is sufficiently constant so that it may be used as a reliable and steady clock for most practical purposes.

The length of the day is defined in terms of the axial rotation of the earth relative to the sun. The length of the year is defined in terms of the revolution of the earth in its orbit about the sun. From the standpoint of establishing a calendar, the problems are how to divide the year into days and how to keep the seasons from shifting their places in the calendar with time.

The rotation of the earth. The simplest way of demonstrating the rotation of the earth is with the Foucault pendulum. In 1851 the French physicist Foucault suspended a heavy iron ball at the end of a 200-foot wire, drew it slightly aside, and allowed it to start swinging as a pendulum. The motion of the pendulum was recorded by means of a pin set in the bottom of the ball, which traced its path in a tray of fine sand. After a while it was obvious that the plane in which the pendulum oscillated was rotating in a clockwise sense, indicating that the earth was rotating

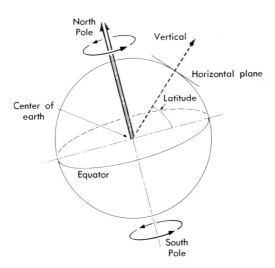

Fig. 8. Rotation of the earth. The earth rotates counterclockwise as viewed from above the North Pole. The double arrow through the North Pole represents the total rotational (angular) velocity of the earth about its axis. The dotted arrow perpendicular to the horizontal (tangent) plane represents the "vertical component" of the earth's rotation (i.e., the rotation of the horizontal plane) at any latitude.

under the pendulum in a counterclockwise sense. (Had Foucault conducted his experiment in the Southern Hemisphere, he would have observed a rotation in the opposite sense.)

The rotation measured by the Foucault pendulum is the vertical component of the earth's total rotation about its axis, as illustrated in Fig. 8, and represents the rotation of the horizontal plane. At the poles, the Foucault pendulum measures the total earth rotation. Elsewhere, it measures the rotation of the horizontal plane, which depends on the latitude.* (At the Equator, the Foucault pendulum does not rotate at all.)

There are slight variations taking place in the rotation of the earth, both regular and irregular, due to the effects of the atmosphere, the distribution of snow and ice, and the tidal motions of the oceans. The slowing down of the earth's rotation due to tidal friction is causing the days to lengthen at the rate of about one-thousandth of a second per century. However, for most purposes the earth is a sufficiently constant clock.

Solar and sidereal days. The passage of an astronomical body across the meridian (longitude circle)† of the observer is referred to as a transit. The time interval between successive transits of the meridian by the sun is called an apparent solar day. (The time interval between successive transits of the meridian by a star is called a sidereal day.) Apparent local noon is the time of transit of the sun.

The length of the apparent solar day varies slightly during the year. The longest apparent solar day, about December 25, is approximately 50 seconds longer than the shortest apparent solar day, about September 13. The cumulative effect of these variations in the length of the day causes the sun alternately to run ahead of or lag behind our clocks by as much as 15 minutes. The variations in the length of the day are due to two effects: the eccentricity of the earth's orbit and the tilt of the earth's axis of rotation. The effect of eccentricity is that the earth advances faster and farther along its orbit near perihelion than near aphelion. As a result, the earth must turn farther for the sun to repeat a meridian transit near perihelion than near aphelion, and this tends to make perihelion days longer. A more important factor is the tilt of the earth's axis, which is discussed in more detail later. The earth's Equator‡ is tilted at an angle of about $23\frac{1}{2}$ degrees relative to the plane of the ecliptic. This angle is called the obliquity of the ecliptic. Because of the obliquity of the ecliptic,

---

* The period of rotation of the Foucault pendulum, i.e. the time required for the plane of oscillation to make one complete revolution, is called the *pendulum day*. The pendulum day is equal to a sidereal day (23 hours 56 minutes) divided by the sine of the latitude.

† Latitude and longitude will be discussed in Section 4.

‡ The equatorial plane is perpendicular to the axis of rotation. The tilt of the earth may be expressed in terms of either Equator or axis.

the sun appears to move during the year along a path through the stars, called the ecliptic, which is inclined with respect to the Equator. The motion of the sun along the ecliptic, rather than along the Equator, causes a periodic variation in the length of the apparent solar day. Because of the tilt of the earth's axis, the "latitude of the sun" (see *declination,* Fig. 16) changes from one solar transit to the next. This north-south "motion" of the sun varies with the season, and is primarily responsible for the variation in the length of the day. The obliquity effect accounts for the phenomenon that the longest apparent solar days occur not at perihelion but at the winter and summer solstices, while the shortest days occur not at aphelion but at the equinoxes.

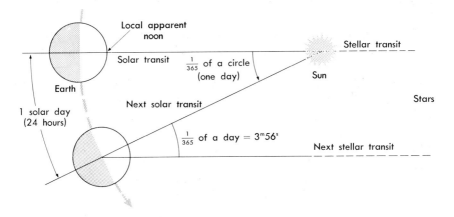

Fig. 9. Difference between solar and sidereal days. The earth sweeps out 1/365 of its orbital circle each day. The earth rotates through the same part of a circle in passing from the stellar transit which ends the sidereal day to the solar transit which ends the solar day. The full rotation of the earth takes 24 hours, so that 1/365 of a rotation takes 3 minutes 56 seconds. This is the difference between the solar and sidereal days.

If the earth's orbit were a circle, and the earth's axis were perpendicular to the orbital plane, the sun would move at a constant rate along the celestial equator, and the length of the day would be constant. This day, which is equal in length to the average apparent solar day, is called the mean solar day. The mean solar day is used as the basis for clock time. By definition, it is divided into 24 hours which, also by definition, are further subdivided into minutes and seconds.

The mean solar day is exactly 24 hours long, whereas the sidereal day is about 4 minutes shorter. The reason for the difference is explained in Fig. 9. The revolution of the earth about the sun causes a longer interval to elapse between successive solar transits than between successive stellar transits.

Apparent, mean, standard, and Greenwich time. If we use the sun to tell time, as with a sundial, we measure apparent solar time. Thus, for example, the sun will transit the meridian (reach maximum elevation above the horizon and cast the shortest shadow) at noon apparent solar time (1200 ast).* The mean solar time, from which we can determine clock time, differs from the apparent solar time by an amount which varies with season. The difference, mean solar time (mst) minus apparent solar time (ast), is called the *equation of time*. (An approximate and abbreviated table of the equation of time is given in the instructions for the laboratory exercise on the sundial. See also the supplementary laboratory notes on the sundial.) To determine mean solar time from apparent solar time (sundial time), we add the equation of time to the latter.

The earth rotates from west to east, causing the apparent motion of the sun across the sky from east to west. The earth rotates through 360 degrees (one complete revolution or a full circle) in 24 hours, or 15 degrees every hour. As a result, the sun traverses 15 degrees of longitude from east to west every hour, or one minute of longitude every four seconds. Thus the local apparent solar time at any moment changes by one hour for every 15 degrees of longitude. The time decreases (is earlier) as one moves westward, in the direction of motion of the sun, and increases (is later) as one moves eastward, opposite to the direction of the sun's motion. By international agreement, the day begins at about longitude 180°, which lies in the Pacific Ocean. This is the international date line. When crossing the international date line from east to west, one "loses" (goes forward) a day, and when crossing it in the opposite direction, one "gains" (goes back) a day. For example, if the time is 1200 ast, February 15 at longitude zero, the longitude of the meridian through Greenwich, England, it is midnight (2400) ast February 15 on the west side of the 180° meridian. However, on the east side of the 180° meridian it is midnight (2400) ast February *14* or 0000 ast February 15. Thus the day changes as we cross the date line.

For the purposes of common timekeeping it is necessary to define time zones within which all clocks read the same time. The time zones are approximately 15 degrees of longitude wide. The central meridian of the time zone is used to define the *standard* time within the zone. These meridians are exactly 15 degrees of longitude, i.e., one hour, apart. The standard time of each zone is equal to the mean solar time on the central

---

* We shall express time in the 24-hour clock system. The first two digits are the hours, from 00 to 24, and the last two digits are the minutes from 00 to 59. Thus 6:30 a.m. will be written 0630, and 6:30 p.m. will be written 1830. Midnight is either 0000 or 2400, the former being used for the beginning of the day, and the latter for the end of the day.

TABLE 3
UNITED STATES TIME ZONES

Standard time	Central meridian
Eastern Standard Time (EST)	75°W
Central Standard Time (CST)	90°W
Mountain Standard Time (MST)	105°W
Pacific Standard Time (PST)	120°W

meridian of the zone. In the United States, the four time zones are defined as shown in Table 3.

It is convenient, especially for geophysical purposes, to have a universal system of time which does not depend on locality. The time commonly employed for this purpose is Greenwich Civil Time (GCT), the mean solar time on the Greenwich (0°) meridian. Greenwich time differs from any other standard time by one hour for every 15 degrees of longitude. For example, when the time is 1200 EST, it is five hours ($\frac{75}{15}$) *later* at Greenwich, or 1700 GCT. When the time is 1300 mst on the 30 degree east meridian, it is two hours *earlier* at Greenwich, or 1100 GCT.*

To determine standard (clock) time from the mean solar time at any locality, we must account for the difference in longitude between the central meridian of the time zone and the longitude of the locality. If the locality is *east* of the central meridian, the standard time (mst on the central meridian) is *earlier* than the local mst. To determine the standard time we therefore *subtract* four seconds from the local mst for every minute of longitude difference between the locality and the central meridian. If the locality is *west* of the central meridian, the standard time is *later* than local

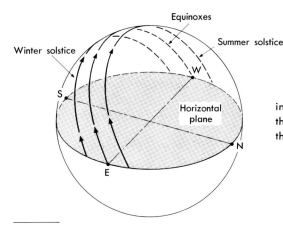

Fig. 10. Seasonal variation in the path of the sun across the sky in middle latitudes of the Northern Hemisphere.

---

* Mean solar time on the Greenwich meridian is also known as Greenwich Mean Time (GMT) and as Universal Time (UT).

mst, and we *add* four seconds to the local mst for every minute of longitude difference to obtain standard time.

The obliquity of the ecliptic, the seasons, and the calendar. The path of the sun across the sky changes with the season, as illustrated in Fig. 10. In winter, in middle latitudes, the sun rises south of east and sets south of west, the days are short, and the elevation of the sun above the horizon is low. In summer, in middle latitudes, the sun rises north of east and sets north of west, the days are long, and the elevation of the sun above the horizon is high. The reason for this seasonal variation is the tilt of the earth's axis, i.e., the obliquity of the ecliptic.

The axis of rotation of the earth is tilted at an angle of 23.5 degrees from the perpendicular to the plane of the ecliptic. Except for a very slow wobbling motion of the axis, called *precession,* the orientation of the earth's axis remains fixed in space as the earth revolves about the sun. The effect of this tilt is illustrated in Fig. 11.

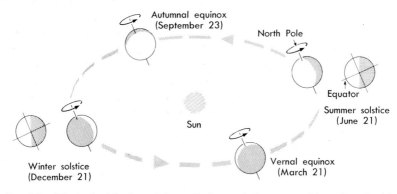

Fig. 11. Effect of obliquity of the ecliptic on the seasons. Note that the North Pole, which is tilted always in the same direction in space, tilts toward the sun at summer solstice, and away from the sun at winter solstice.

At the time of the summer solstice, which occurs about June 21, the sun appears directly overhead at noon on latitude 23.5 degrees north, the Tropic of Cancer. The sun is farthest north in its annual migration at this time, and "pauses" before beginning its southward march. Hence the name solstice, from the Latin word for "sun stand still." The elevation of the sun above the horizon and the length of the day reach their maximum values in the Northern Hemisphere at the summer solstice, and everywhere north of the Arctic Circle, latitude 66.5 degrees north, the sun remains above the horizon all day. In the Southern Hemisphere, the sun's elevation is at a minimum, the days are shortest, and everywhere south of the Antarctic Circle, latitude 66.5 degrees south, the sun does not rise above the horizon on the June solstice. This is the beginning of the Northern Hemisphere summer and the Southern Hemisphere winter.

Winter in the Northern Hemisphere, and summer in the Southern Hemisphere, begin at the winter solstice, about December 21. Then the sun, having reached the southernmost point in its annual migration, stands directly overhead at noon on latitude 23.5 degrees south, the Tropic of Capricorn. Both the elevation of the sun above the horizon and the length of the day are then at their minimum values in the Northern Hemisphere (maximum in the Southern Hemisphere), and the sun does not rise within the Arctic Circle or set within the Antarctic Circle. At the vernal (spring) and autumnal equinoxes, the days and nights are everywhere equal (12 hours), and the sun appears directly overhead at noon on the Equator. The sun crosses the Equator from north to south at the autumnal equinox, and from south to north at the vernal equinox. The ecliptic (the apparent path of the sun among the stars) intersects the celestial equator at the equinoxes.

The time interval between successive vernal equinoxes is called the tropical year, and is equal to 365 days 5 hours 48 minutes and 45.7 seconds. The tropical year is slightly shorter than the sidereal year because the equinoxes slowly shift their positions in the earth's orbit. This effect, which is known as the precession of the equinoxes, is caused by the slow wobbling motion of the earth's axis. (The axis completes its circle of precession every 25,800 years. The precession of the axis also changes the position of the North Star, which at the moment is Polaris.) To keep the seasons always in the same months in the calendar, the tropical year is used as the basis for the calendar.

The Julian calendar, established by Julius Caesar in 45 B.C., contained 365 days, and 366 every fourth year (leap year) to take care of the extra 5 hours 49 minutes. Thus the average length of the Julian year was 365 days 6 hours, or about 11 minutes longer than the tropical year. The accumulation of this discrepancy over a period of many centuries eventually caused the vernal equinox to move into the winter months. The calendar was corrected in the sixteenth century by edict of Pope Gregory. In the first place, the date March 21 was fixed to correspond with the vernal equinox. Then the leap years were eliminated in the years 1700, 1800, and 1900 A.D. (but not in 2000 A.D.). This shortened the average year by almost the required amount, but not quite. The Gregorian calendar year is still about 26 seconds longer than the tropical year and will have to be adjusted again some day.*

4. The shape and size of the earth. Geodesy is the science of measurements on the earth. It is concerned with the measurement of the earth itself, as well as with measurements of portions of the earth's surface.

---

* See references 11 and 13.

Geodesic measurements of the earth have been made by astronomical and gravimetric (gravity) methods. (We shall consider the earth's gravity and its variations, and explain how gravity measurements lead to information about the shape of the earth during our discussion of gravity.) Geodesic measurements on the earth's surface, for mapping and other purposes for which precise distance and position data are required, are carried out by the surveying technique known as triangulation.

Observations by means of artificial earth satellites have, since 1957, enabled geodesists to compute the dimensions of the earth more precisely than ever before. Satellite determinations of the earth's shape and size are based on photographic measurements (photogrammetry) of the earth, as well as on gravity calculations derived from satellite orbits.

History of geodesy. In the sixth century B.C., the Greek philosopher Pythagoras (*ca.* 582–500 B.C.) taught that the earth was a sphere. This idea was apparently based on the phases of the moon, which led Pythagoras to deduce that the moon must be a sphere, and hence that the earth is probably a sphere also. About 200 years later, Aristotle (384–322 B.C.) was led to the same conclusion by the circular form of the shadow that the earth cast on the moon during a lunar eclipse. In the third century B.C., Eratosthenes (*ca.* 284–192 B.C.) not only demonstrated the spherical shape of the earth, but actually measured the radius of the sphere with astonishing accuracy.

The next significant advance in geodesy came in 1617 when Snell precisely calculated the circumference of the earth from astronomical measurements. Then in 1671, while on an expedition to French Guiana, Richer discovered that gravity was apparently weaker near the Equator than in France. From this he concluded that the earth could not be a perfect sphere. Isaac Newton, in 1686, finally gave the correct description of the shape of the earth. He described it as an oblate spheroid, with its equatorial radius longer than the polar radius, computed the ellipticity of the spheroid, and explained the flattened shape of the earth as being caused by its rotation. The final proof and calculations of the oblate spheroidal shape of the earth were given by Clairaut in 1743 on the basis of theoretical calculations and measurements of gravity. Careful geodetic measurements since the eighteenth century have necessitated only minor changes in Clairaut's model of the earth. Recent satellite measurements have resulted in two small modifications in the shape of the earth model. One was the discovery that the Southern Hemisphere has a slightly larger circumference than corresponding latitudes in the Northern Hemisphere, making the earth somewhat "pear-shaped." The other was the observation that the equatorial cross section of the earth is not an exact circle, but appears to be slightly elliptical.

The dimensions of the earth. The calculation of Eratosthenes is illustrated in Fig. 12. Eratosthenes determined that on the longest day of the

CHAP. 1]   THE SHAPE AND SIZE OF THE EARTH   23

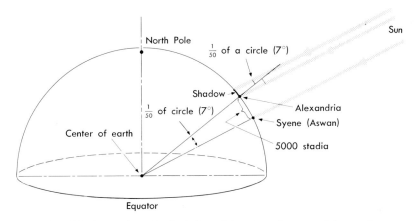

Fig. 12. Eratosthenes' calculation of the circumference of the earth.

year (the summer solstice), the sun appeared directly overhead at noon at the ancient city of Syene (now called Aswan) in Egypt. On the same day, at noon, a pole at Alexandria cast a shadow on the ground whose length indicated that the zenith angle of the sun (the angle between the sun and the vertical) was one-fiftieth part of a circle (7°). As shown in the figure, the angle subtended by the radii from the center of the earth to Alexandria and to Syene must also be one-fiftieth of a circle. The distance from Syene to Alexandria was known to be 5000 stadia,* or about 925 kilometers. Thus Eratosthenes calculated that the circumference of the earth must be 50 × 5000, or 250,000, stadia, which is about 46,000 kilometers (29,000 miles). The radius of the earth can easily be calculated from the circumference,† and was found to be 7400 kilometers (4600 miles). This result is only about 15 percent higher than the correct value.

The rotation of the earth about its axis has caused the earth to assume the shape of an oblate spheroid, i.e., of a ball flattened slightly at the poles, so that the polar radius, the distance from the center of the earth to the pole, is shorter than the equatorial radius, the distance from the center of the earth to any point on the Equator. Because of the distribution of land masses and mountains, the earth is a rather "warty" looking spheroid. For reference purposes it is convenient to postulate a smooth

---

* There is some doubt about the precise definition of the ancient length unit, the "stadium." However many scholars believe that 1 stadium was equal to about 185 meters or 605 feet. Thus 5000 stadia would equal 925,000 meters or 925 kilometers. (The conversion from meters to feet is 1 foot = 0.305 meter, and from kilometers to miles, 1 mile = 1.609 kilometers.)

† The ratio of the circumference of a circle to its diameter (twice the radius) is equal to 3.1416 ... and is known as π (the Greek letter *pi*).

earth with about the same over-all dimensions as the real earth. The international reference spheroid adopted by the International Union of Geodesy and Geophysics (IUGG) has approximately the following dimensions:

Polar radius: 6357 km (3951 mi),
Equatorial radius: 6378 km (3964 mi),
Mean radius: 6371 km (3960 mi).

The international reference spheroid differs slightly from another reference surface employed in geodesy. This is the geoid, which represents a *level surface,* i.e. one that is perpendicular to the direction of the force of gravity. To visualize the geoid, imagine the surface of the earth to be completely covered with water. The surface of this endless sea (sea level) would correspond to the geoid. It, too, is an oblate spheroid which departs slightly from the reference spheroid because of irregularities in the distribution of the earth's mass. For the remainder of our discussion we shall ignore the oblateness of the earth and assume it to be a perfect sphere with a radius of 6371 kilometers (3960 miles).

The surface area of the earth is easily calculated from its radius,* and is found to be about 198 million square miles. Similarly, the volume may also be calculated from the radius, and is found to be equal to about 260 billion cubic miles or 1.08 million million cubic kilometers. We shall see shortly that the total mass of the earth may be determined from the gravitational force which it exerts, and is found to be almost 6×10^{24} kilograms.† The average density of the earth is calculated by dividing its mass (5.98×10^{27} grams) by its volume (1.08×10^{27} cubic centimeters), from which we obtain 5.52 grams per cubic centimeter. (The density of pure water is 1 gram per cubic centimeter.)

Longitude and latitude. The meridians, or longitude lines, on the earth are great circles passing through the North and South Poles. A great circle is constructed by passing a plane through the center of the earth. The intersection of such a plane with the spherical surface of the earth is called a great circle. A great circle is the largest circle one can draw on the surface of a sphere. The shortest distance between any two points on

---

* The formula for the surface area of a sphere is 4π times the square of the radius ($4\pi r^2$). The formula for the volume of a sphere is $4\pi/3$ times the cube of the radius ($4\pi r^3/3$).

† One kilogram (kgm) is equal to about 2.2 pounds. One thousand kilograms (2200 pounds) is called a metric ton, and is slightly larger than the ordinary (English) ton (2000 pounds). The mass of the earth is about 6×10^{21} metric tons. (The notation 10^{21} denotes a one followed by 21 zeros, and is a convenient way of writing large numbers. We will employ this "exponential" system from now on. In the exponential notation one million is written 10^6, one billion 10^9, etc.)

the surface of a sphere is along the great circle between them. Great circles are also called geodesics.

The length of a meridian circle, i.e., the circumference of the earth measured around the poles, is approximately 40,000 kilometers (24,900 miles). The distance from Equator to pole along a meridian is one-fourth of the circumference, i.e., approximately 10,000 kilometers (10 million meters). When the metric system was adopted, it was intended that the standard unit of length, the meter, should be about one ten-millionth of the distance from Equator to pole.

The meridians are labeled in degrees of longitude from zero (the meridian passing through Greenwich, England) to 180 degrees. Meridians west of Greenwich are labeled in degrees west, while those east of Greenwich are labeled in degrees east. There are 360 degrees of longitude around the earth. One can determine the longitude of any one place by observing the sun—if one has an accurate clock (chronometer). For example, when the sun transits the local meridian, the time is 1200 apparent solar time. From the equation of time and the date we can convert apparent solar time to mean solar time. For every 15 degrees of longitude west (east) of Greenwich, the local mean solar time is one hour earlier (later) than Greenwich Civil Time (GCT). Thus, if our chronometer reads Greenwich time, we can determine our longitude by multiplying the difference between GCT and local mean solar time by 15 degrees of longitude per hour. (To cite an example, let us assume that at 1200 local mean solar time our GCT chronometer reads 1500 GCT. The time difference is plus 3 hours, and our longitude is therefore 45 degrees west.)

The latitude circles, or parallels, are concentric circles on the surface of the earth, constructed by passing parallel planes through the sphere at right angles to the axis of rotation. The intersections of the planes with the sphere are the latitude circles. Parallels are labeled in degrees of latitude, from zero at the Equator (the only one of the latitude circles that is also a great circle) to 90 degrees at the poles. In the Northern Hemisphere, latitudes are labeled in degrees north, while in the Southern Hemisphere they are labeled in degrees south.

There are three kinds of latitude. *Astronomical latitude,* illustrated in Fig. 13, is defined as the angle between the direction of a plumb line and the plane of the Equator. A plumb line consists of a weight hanging at the end of a string. The plumb line indicates the direction of gravity and, by definition, the vertical direction. *Geocentric latitude* is the angle between a radius drawn from the center of the earth to a point on the earth's surface and the equatorial plane, as shown in Fig. 14. If the earth were a stationary, uniform sphere, there would be no difference between astronomical and geocentric latitude, for gravity would then be directed everywhere toward the center of the earth. The latitude used in the construction of maps is called *geographic latitude.* This is the astronomical

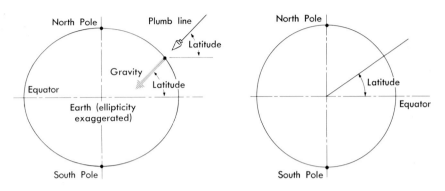

Fig. 13. Astronomical latitude.   Fig. 14. Geocentric latitude.

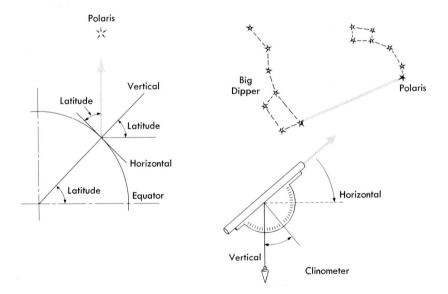

Fig. 15. Determination of latitude from observation of Polaris.

latitude corrected for the so-called "station error," i.e., local gravity anomalies caused by nonuniform distribution of the earth's mass. Lines of equal astronomical latitude would be slightly wiggly because of these irregularities, whereas the geographic latitude circles are perfect circles. The nearly spherical form of the earth is indicated by the fact that the maximum difference between geographic and geocentric latitude is less than 12 minutes of arc.

Two methods for determining your latitude (approximately) are illustrated in Figs. 15 and 16. In the first method, the latitude is measured by measuring the elevation of the north star, Polaris, above the horizon.

Polaris may be located in the northern sky by reference to Ursa Major, the Big Dipper, as shown in Fig. 15. The line of sight to Polaris is nearly parallel to the axis of the earth, and the elevation of this line above the horizon is equal to the latitude. One may construct a clinometer for measuring elevation angles above the horizon with a drinking straw, a protractor, a string, and a weight, as illustrated in the figure.

A second method for measuring latitude, presented in Fig. 16, proceeds by observing the elevation of the sun above the horizon at local apparent noon (when the sun is on the meridian). The angle of elevation of the sun *above the equatorial plane* is called the declination. It varies with season from $+23\frac{1}{2}$ degrees on the summer solstice, through zero at the equinoxes, to $-23\frac{1}{2}$ degrees on the winter solstice. The declination for any date can be found in astronomical tables (*American Ephemeris and Nautical Almanac*). The angle between the vertical and the sun is called the zenith angle of the sun, and is equal to 90 degrees minus the elevation angle of the sun above the horizon. From the figure, we can see that the latitude is equal to the sum of the zenith angle (measured) and the declination (determined from the date and tables).

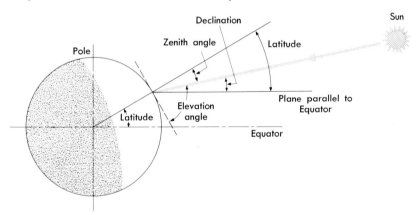

Fig. 16. Determination of latitude from elevation of the sun at noon.

The meridians and parallels intersect at right angles on the surface of the earth. The length of one degree of latitude along a meridian is equal to about 111 kilometers (69 statute miles); it is also equal to 60 nautical miles, for a nautical mile is defined as approximately one minute of latitude and thus is a convenient scale for measuring distance on maps. As there are 60 minutes in a degree of latitude, one nautical mile corresponds to one minute of latitude.*

---

* See references 4 and 11.

5. Gravity. The concept of gravity is one of the fundamental cornerstones of classical physics. It has enabled scientists to explain the mechanics of the solar system and to predict the trajectories of projectiles on the earth. It explains the phenomenon of weight, the acceleration of falling objects, and the orbits of earth satellites. Gravity is universal, and enters into all branches of physics and all aspects of life.

In geophysics, gravity is of special interest for several reasons. First of all, gravity measurements have been used to determine the mass of the earth. Secondly, these measurements lead to information about the shape of the earth. Careful observations of gravity also provide us with data about the distribution of matter under the earth's surface. As a practical application of these gravity measurements, it is possible to locate oil deposits and other irregularities in the composition of the earth's crust.

Although gravity is nearly constant over the surface of the earth, it varies slightly and systematically with latitude. In addition, there are also small-scale variations caused by irregularities in the earth's mass (e.g., mountains). Accurate gravity meters (gravimeters) are used to measure and map these gravity anomalies.

Historical review. Aristotle believed that the speed of a falling body depended on its mass. In the early seventeenth century, Galileo refuted this Aristotelian theory, demonstrating with the aid of balls rolling down an inclined plane that all bodies fall at the same rate, regardless of their mass. Specifically, Galileo showed that all bodies fall with the same constant acceleration. The acceleration of a body is the rate at which its velocity changes with time. The acceleration of a falling body near the earth's surface is found to be about 980 centimeters per second per second (980 cm·sec^{-2}) or 32 feet per second per second. This means that the speed of a falling body increases by 980 centimeters per second every second. (Actually this result is true only in a vacuum where the falling object is not slowed down by air resistance. Also, as we shall see, this acceleration is not the same everywhere on the earth. Neither of these facts diminishes the great importance of Galileo's discovery.)

More than a half-century later, Newton formulated the law of universal gravitation to explain the motions of the planets and the moon. According to Newton, the planets are attracted to the sun by a force, gravitation, which exists by virtue of the masses. The magnitude of the force is proportional to the masses of the sun and the planet and inversely proportional to the square of the distance between them.* The planets "fall" toward the sun as a consequence of this mass attraction. Thus they are pulled inward, away from the direction of a straight line in space, and constrained to move

---

* The "distance between bodies" in the law of gravitation is the distance between the centers of mass of the bodies.

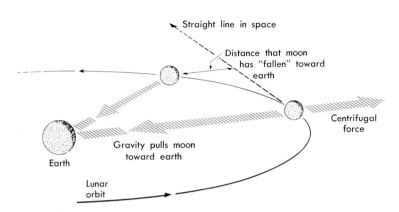

Fig. 17. The centripetal motion of the moon caused by gravitational attraction between moon and earth.

in curved (elliptical) paths about the sun. Newton proved his law by calculating the gravitational effect of the earth on the moon from the orbit of the moon, as illustrated in Fig. 17.

The gravitational force causes the moon (planet) to accelerate constantly toward the earth (sun) as it moves in its orbit. This is called a centripetal acceleration. Because of inertia, the revolving object would move in a straight line, if it were not for gravity. The tendency for the body to move in a straight line may be represented by an imaginary outward-directed force, called the centrifugal force. We may think of gravity as opposing this inertial force. By imagining a centrifugal force, we are perhaps helped to see how the revolving body stays in its orbit. The notion of a centrifugal force is also convenient for purposes of making calculations. (The centrifugal force is equal to the mass of the revolving body times its radial distance from the center of rotation times the square of its angular velocity.)

Together with Newton's second law of motion, stating that the acceleration of a body is equal to the force applied divided by the mass of the body, the law of gravitation explained Galileo's law of falling bodies. A falling body on earth is pulled down by the force of gravity, which is proportional to the mass of the body times the mass of the earth (a constant), and inversely proportional to the square of the distance between the center of mass of the body and the center of mass of the earth. Bodies near the earth's surface are at nearly constant distance (the radius of the earth) from the earth's center. Thus the gravitational *force* is simply proportional to the mass of the falling body. But the *acceleration* resulting from the force of gravity, which is what Galileo measured, is equal to the force divided by the mass. Hence the mass is cancelled, and the gravitational *acceleration* does not depend on the mass.

Newton not only explained why Galileo found that the acceleration of gravity is constant, but he also verified Richer's discovery that gravity varies with latitude. Later, in the eighteenth century, precise measurements of gravity led to the accurate determination of the shape of the earth. At the same time, variations in the force of gravity were observed and related to mountains and to irregularities in earth density. An important discovery based on gravity measurements is the fact that mountains apparently "float" on relatively light underlying material, a phenomenon known as *isostasy*.

To complete Newton's work, it was necessary to determine the constant of proportionality (the universal gravitational constant) appearing in his law of universal gravitation. This could be done by measuring the mass of the earth or, in a laboratory, by measuring the gravitational force between two known masses a fixed distance apart. Since we know the radius of the earth and the acceleration due to gravity at the earth's surface, we need know only the earth's mass to compute the constant of proportionality in the formula. Conversely, if we know the constant of proportionality, we can calculate the earth's mass from the gravitational acceleration. Having determined the earth's mass, we can then use the law of gravitation to determine the masses of the sun, the moon, and the planets.

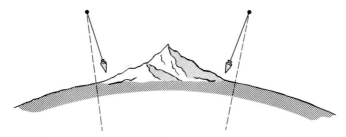

Fig. 18. Deflection of a plumb line by the mass of a mountain (Maskelyne, 1774). (The effect of isostasy is neglected.)

In 1738, Bouguer tried unsuccessfully to measure the gravity constant by measuring the deflection of a plumb line caused by the gravitational pull of a mountain in Peru, the mass of the mountain being known from borings. Maskelyne, in Scotland in 1774, was more successful in using the mountain method, which is illustrated in Fig. 18. The deflection of the plumb line from an astronomical "vertical" was measured on two sides of a mountain. The mass of the mountain, which causes a horizontal gravitational attraction and resultant deflection of the plumb line, was measured independently from rock samples and the mountain's volume, so that the gravitational constant and the mass of the earth could be determined. (It was later found that the plumb line actually may be deflected away from the mountain due to the effect of isostasy. In fact,

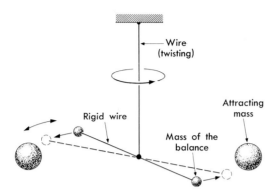

Fig. 19. Use of the torsion balance to measure the universal gravitational constant (Cavendish, 1798).

Bouguer was surprised to find that his mountain caused a smaller deflection effect than he had expected, and he was thus the first to discern the phenomenon of isostasy.)

The first accurate measurement of the gravitational constant was made by Henry Cavendish (1731–1810) about 1798, following a suggestion by J. Michell. Cavendish used a torsion balance and known masses to measure the force of gravity, as illustrated in Fig. 19. By measuring the oscillations resulting from deflection of the masses attached to a suspended wire when two larger attracting masses were placed near them, Cavendish was able to measure the gravitational force between the attracting bodies. Since the masses and the distances between them were known, he was able to compute the universal gravitational constant.

Weight, mass, and acceleration. The force of attraction between the earth and any object in the gravitational field of the earth is called the *weight* of the object. The weight of an object is equal to the product of its mass times the acceleration of gravity. If we carry an object from one place on the earth to another place where gravity is different, its mass does not change, but its weight does. We can measure gravity by measuring the weight of a known mass.

The acceleration which the force of gravity imparts to a body falling in a vacuum is called gravitational acceleration. Although the force of gravity increases with the mass of the falling object, the acceleration, being equal to the force divided by mass, does not. The acceleration due to gravity is the gravitational force per unit mass. The average gravitational acceleration on the earth's surface is about 980 cm·sec^{-2}.

The gravitational force (weight of a body) is proportional to the mass of the body times the mass of the earth divided by the square of the distance from the body to the center (of mass) of the earth. The farther a mass is from the mass center of the earth, the smaller is the acceleration of gravity, and the smaller is its weight. Thus gravity and weight decrease with increasing altitude above the earth. A body at the North Pole is closer to the center of mass of the earth than is a body on the Equator, and hence the gravitational force is greater at the poles than at the Equator. This phenomenon is one of the reasons explaining the observation that objects weigh more at the poles than at the Equator. (Another reason is the rotation of the earth, which is discussed later.)

The gravitational force on the surface of any object, say the moon, may be compared with that on the earth by means of the law of gravitation. The ratio of the forces is proportional to the ratio of the masses, and inversely proportional to the ratio of the squares of the radii. Thus from Table 2 we find that gravity on the moon is only 17 percent of that on earth. If you weigh 100 pounds on earth, you would weigh only 17 pounds on the moon.

The universal gravitational constant of Newton's law, as determined by Cavendish in 1798, is now known to be 6.67×10^{-8} cm$^3 \cdot$gm$^{-1} \cdot$sec^{-2}, a number you need not remember. From this number we can calculate the mass of the earth, if we know the earth's radius (which was measured before its mass) and the acceleration of gravity (measured by Galileo). This is how the mass of the earth, mentioned earlier, was determined.

The variation of gravity with latitude. Imagine a soft plastic sphere rotating about its axis. The effect of the centrifugal force (inertia) is to push the mass outward from the axis of rotation. The centrifugal force increases in proportion to the distance from the axis of rotation. As a result, the sphere bulges outward around the Equator, and an oblate spheroid, like the earth, results.

Any object on the earth is rotating with the earth's surface, and is therefore subject to an outward-directed centrifugal force, which is perpendicular to the axis of rotation of the earth (Fig. 20) and increases from zero at the pole to a maximum at the Equator. This variation occurs because the centrifugal force increases with the square of linear speed, and an object on the Equator is traveling at maximum speed, while one at the pole is not moving at all (except with the earth as a whole).

The gravitational force due to the earth's mass is directed toward the center of the earth (almost) everywhere (see Fig. 20). The *measured* gravity of the earth is the resultant of the combination of mass gravitation and centrifugal force. These two forces and the resultant gravity are illustrated by the arrows in Fig. 20. It can be seen that because of the centrifugal force, gravity must be weaker at the Equator than at the poles. Gravity is about 5 cm·sec^{-2}, i.e. approximately 0.5 percent, greater at the poles than at the Equator.

Not only is the magnitude of gravity variable over the earth, but so is the direction. As seen in Fig. 20, resultant (measured) gravity is not geocentric. The shape of the earth can be represented approximately by

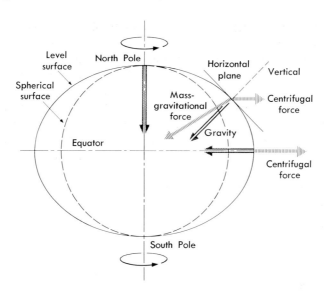

Fig. 20. Effect of centrifugal force of earth's rotation on gravity and the shape of the earth. The double arrows represent measured gravity, the resultant of the gravitational force due to the earth's mass (directed toward the center of the earth) and the centrifugal force (directed perpendicular to the axis of rotation).

a surface (the geoid) which is everywhere at right angles to the local direction of gravity. This surface corresponds closely to sea level, for the level of the sea adjusts itself to the field of gravity. (If gravity were not perpendicular to the level of the sea, the water would "roll downhill" in the direction of the "horizontal" component of gravity.) A level surface is, by definition, one that is perpendicular to gravity. Similarly a vertical line is, by definition, in the direction of gravity. The level surface departs slightly from the spherical shape, and the vertical lines are not exactly lines radiating from the earth's center. The latter effect accounts for the difference between geocentric and astronomical (gravimetric) latitude.

Measurement of gravity. The acceleration due to gravity may be measured by any of three methods. We may measure the acceleration of a falling object, as Galileo did, although this is not practical for routine field measurements of gravity. We may measure the weight of a standard mass by attaching it to a wire or spring and observing the deflection, up or down, of the mass as we move it from place to place. A device of this sort is known as a spring-type gravimeter, and is widely used for gravity measurements. The third method of measuring gravity uses a pendulum.

The principle of the pendulum is illustrated in Fig. 21. A simple pendulum swings back and forth because the mass of the pendulum is pulled down by gravity. The stronger the pull of gravity, the faster the pendulum swings. It can be shown theoretically that the period of a pendulum (the time required for the pendulum to swing through one complete cycle) depends only on the length of the pendulum and on gravity. The period of a pendulum is therefore constant at any given location, and can be used to regulate a clock. However, if the pendulum is moved to a place (e.g., the Equator) where gravity is weaker, it will swing more slowly, its period will be longer, and the clock will "run slow." This is how Richer discovered the variation of gravity with latitude. We may use the pendulum to calculate gravity from observations of the period of the pendulum. (See Lab. Ex. 2.) Some accurate gravimeters employ pendulums.

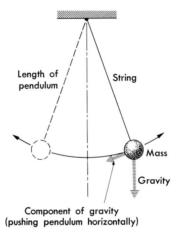

Fig. 21. Principle of the measurement of gravity with a pendulum.

Isostatic equilibrium. Bouguer's observation that a mountain does not deflect a plumb line as much as expected gave rise to the suspicion that the density of the earth may be smaller under mountains than elsewhere. Careful measurements of gravity on tops of mountains and over flat terrain or over water (see Fig. 22) verified this suspicion. Despite the greater volume of earth represented by a mountain, gravity is no stronger on mountain tops than at the same height over water or flat land. In fact, gravity is less than average on mountains, and greater than average over the oceans. Apparently the mountain is "floating" on relatively light underlying material.

It appears that equal sectors of the earth, whether or not they include mountains, contain almost equal masses of earth (except for the latitude effect). This is the phenomenon of isostatic equilibrium, or isostasy.

As we shall see later, the light crustal material of the earth extends deeper under mountains than under the sea. (The mountains have deep, lightweight "roots.") The dense rock of the mantle beneath the crust is

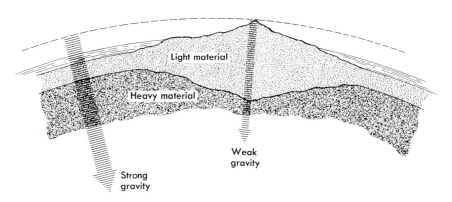

Fig. 22. Isostatic equilibrium. Gravity is stronger over the ocean than over a mountain at the same latitude. The earth is denser under the ocean than under the mountain.

closer to the earth's surface under the sea than under the continents. Thus, while the volume of a column of earth is greater under a mountain than under the sea, the density of the column is greater under the sea.*

6. Maps. A map is a representation on a plane surface of all or a part of the surface of the earth.

Maps are used in all branches of geophysics to display the distribution of geophysical quantities—gravity, geomagnetism, earthquakes, ocean currents, winds, etc.—on the earth. Because different maps have different properties, it is important for the geophysicist to select the correct map for any particular purpose.

It is impossible to "unwrap" the surface of a sphere and flatten it onto a plane. To transform the spherical surface into a plane surface (map), cartographers resort to the device of map projection. The surface of a sphere may be projected on a plane by an (imaginary) optical projection— as if a light passing through the sphere were to cast an image of the earth's surface upon a screen. On the other hand, it is also possible to compute a map projection without resorting to an optical analogy.

No matter how a map is constructed, it is distorted. In the following pages we shall describe several kinds of map projections, and indicate how each of them distorts the appearance of the earth's surface.

The principles of map projection were known to the ancient Greeks. One of the most ancient maps, the gnomonic projection which shows great circles on the earth as straight lines, was apparently invented in the sixth

---

* See references 4, 12, and 14.

century B.C., in the era of Pythagoras and Thales. In the second century B.C., the great Hipparchus invented not only trigonometry but at least two kinds of map projections, the stereographic and orthographic. The stereographic projection is still one of the most useful map projections in the cartographer's file.

In 1538, Gerhard Krämer, known as Mercator, published a map of the world on which areas were represented in true proportion, the first equal-area map of the world. Mercator's most famous projection, the one bearing his name, was constructed in 1569, and is still the most widely used map for long-distance navigation, although it is not an equal-area map.

Important contributions to map projections were made by Lambert, Lagrange, and others in the eighteenth century.

The map projection problem. A spherical surface cannot be "unrolled" onto a plane because the sphere is "generated" by rotating a circle about its diameter. Cones and cylinders, on the other hand, can be flattened into planes by unrolling them because they are "generated" by rotating straight lines.* A map projection of the earth, or any sphere, may be constructed by projecting the spherical surface onto a cone or a cylinder, and then unrolling that surface onto a plane. Or the map projection may be accomplished by projecting directly onto the plane.

One constructs a map by first drawing the meridians and parallels on it, and then adding the geographical features from a knowledge of their latitudes and longitudes. The construction of the meridians and parallels on the map is the concern of map projection, a branch of applied mathematics. The addition of the geographic features is the work of the surveyor, geodesist, geographer, and cartographer.

The objective of a map projection may or may not be to represent the earth's surface with minimum distortion. Ideally, we would like to represent the earth exactly as it is. However, this can be done only with a globe. Since no map can be perfect, we may often allow a map to be greatly distorted for the sake of achieving some other desirable property.

Desirable properties of map projections. For geographical purposes it is desirable that areas on the earth be shown in correct proportion on the map. Equal areas on the sphere should appear as equal areas on the map. A map with this property is called an *equal-area* map.

In geophysics it is often more important that the angles between lines on the earth be truly represented on the map. If a map represents angles correctly, it will also represent the shapes of terrestrial features accurately. A map which preserves angles and shapes is called a *conformal* map. The meridians and parallels on a conformal map intersect at right angles, just as

---

* A surface which can be unrolled into a plane is called a developable surface.

they do on the earth. However, it does not follow that every map on which meridians and parallels intersect at right angles is a conformal map.

The scale of a map is the ratio of distance on the earth to the corresponding distance on the map (e.g., 1000 miles to the inch). It is desirable, but uncommon, for the scale to be constant over the entire map.

Navigators prefer a map on which a straight line drawn from any point to any other point on the map shows the true direction of the line. Such a map is said to be *azimuthal* (from azimuth, an Arabic word referring to directions from north or south).

A curve which crosses successive meridians at a constant angle on the earth is called a loxodromic curve, or a *rhumb line*. This is the course a ship or airplane would follow if it were holding a constant compass heading (corrected for magnetic variation). It is desirable that the rhumb line between any two points on the earth appear as a straight line on the map. However, only one map (the Mercator projection) has this property.

It may also be desirable that great circles appear as straight lines on the map. A map exhibiting this property is called a *great-circle sailing chart*.

Finally, it is convenient to have a map on which the distances between points on the earth are represented in correct proportion on the map. Such a map is said to be an *equidistant* map.

No map possesses all these desiderata. In fact, only a few contain as many as *two* of the desirable features.

Types of map projection. The three general types of map projections are the *zenithal,* in which the sphere is projected directly onto a plane, the *conic,* in which the sphere is projected onto a cone, and the *cylindrical,* where it is projected onto a cylinder.

Among the zenithal, or tangent-plane projections, are the following:

Orthographic. This map is constructed by projecting parallel rays through the sphere onto a plane tangent to the sphere. It is neither equal-area nor conformal.

Stereographic. The stereographic projection illustrated in Fig. 23 is constructed by projecting radiating rays through the sphere from a point on the sphere opposite the tangent plane. In the north-polar stereographic map, the rays emanate from the south pole, the meridians are radiating straight lines comparable to the spokes of a wheel, and the parallels appear as concentric circles. This is one of the few *conformal* maps. Areas near the Equator are greatly exaggerated on the polar stereographic map.

Gnomonic. In this projection the projecting rays emanate from the center of the earth. As a result, all great circles appear as straight lines on the gnomonic map. A polar gnomonic map superficially resembles a stereographic map, for the meridians and parallels are perpendicular to each other. But the gnomonic map is not conformal. It is used principally as a great-circle sailing chart. To find a great-circle (shortest-distance)

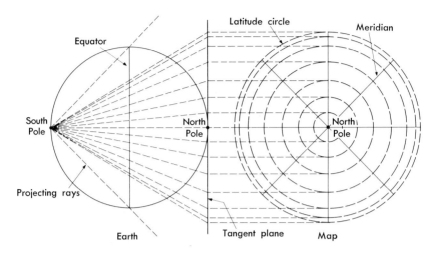

Fig. 23. Polar stereographic map projection.

course between any two points on the gnomonic map, a navigator simply draws a straight line connecting the points.* Areas are greatly distorted near the Equator on a polar gnomonic map. In fact the Equator cannot be shown on the map at all (see Fig. 24).

Other zenithal maps are the azimuthal equidistant and the azimuthal equal-area projections, whose properties are indicated by their names.

All zenithal maps are azimuthal in the sense that the direction of any line passing through the tangent point (the center) of the map is correctly represented.

The construction of a conic projection is illustrated in Fig. 25(a). In this simple conic map projected onto a tangent cone, the map scale printed on the map applies only along the tangent parallel (called the standard parallel). Elsewhere, distances are exaggerated somewhat. This projection is neither equal-area nor conformal.

A conic map projection may be constructed with two standard parallels by passing the cone through the sphere, as shown in Fig. 25(b). Between the standard parallels, the scale of this map is nearly constant. To convert the conic map with two standard parallels into a *conformal* map, it is necessary to adjust the spacing of the parallels mathematically. The resulting map (Fig. 25c), widely used in meteorology, is known as a *Lambert conformal conic* projection, after its inventor.

---

* To find the compass headings that he must hold to remain on the great-circle course, however, the navigator transfers the points on the great-circle course from the gnomonic map to a Mercator map. On the latter, the great circle appears as a curved line, indicating that in general, the compass heading varies continuously along a great circle.

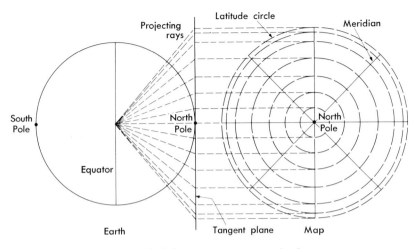

Fig. 24. Polar gnomonic map projection.

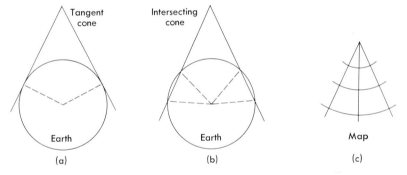

Fig. 25. (a) Simple conic projection. (b) Conic projection with two standard parallels. (c) Conic map.

A popular form of conic projection among geographers is the polyconic, in which a composite of many cones and many standard parallels is used to minimize scale distortion.

Among the cylindrical projections the best known is the *Mercator*. Although this map is actually a computed map, and Mercator himself did not think of it in terms of a cylinder, it can be visualized as follows. Imagine a cylinder wrapped around the earth, tangent to the Equator, with a light source located near the earth's center. Let the light shine through the sphere onto the cylinder, projecting the earth on the cylinder. On the map the meridians will be projected as straight, parallel, equally spaced vertical lines, and the parallels will appear as straight, parallel, unequally spaced horizontal lines. Unroll the cylinder, and make a few mathematical adjustments of the spacing of the parallels, and the result is the Mercator

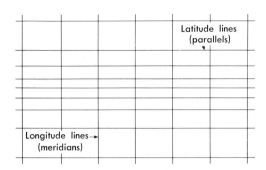

Fig. 26. Mercator map projection.

map (Fig. 26). The Mercator map is conformal. Furthermore, a rhumb line between any two points on the map appears as a straight line. To plot a course, a navigator draws a straight line on the Mercator map between his point of origin and his point of destination, and measures the (azimuth) angle between this rhumb line and the meridian. This is the constant compass heading (corrected for magnetic variation) that he must hold to arrive at his destination. The Mercator map greatly exaggerates areas near the poles and has a highly variable scale.

In this brief review we have omitted the discussion of a large number of specialized maps, including several equal-area maps preferred by geographers.*

---

* See references 13 and 15.

References

1. P. M. Hurley, *How Old Is the Earth?* New York: Science Study Series, Anchor Books, Doubleday, 1959.
2. G. P. Kuiper, "Origin, Age, and Possible Ultimate Fate of the Earth," in *The Earth and Its Atmosphere,* D. R. Bates, ed. New York: Basic Books, 1957.
3. H. C. Urey, "The Origin of the Earth," in *The Planet Earth,* A Scientific American Book. New York: Simon and Schuster, 1957.
4. B. F. Howell, Jr., *Introduction to Geophysics.* New York: McGraw-Hill, 1959.
5. J. A. Jacobs, R. D. Russell, and J. T. Wilson, *Physics and Geology.* New York: McGraw-Hill, 1959.
6. H. Brown, "The Age of the Solar System," *Scientific American,* **196,** 80–104 (April 1957).
7. J. H. Reynolds, "The Age of the Solar System," *Scientific American,* **203,** 171–182 (November 1960).
8. "The Universe," *Scientific American,* **195** (September 1956). (Eleven articles on cosmology and cosmogeny.)
9. K. Krauskopf and A. Beiser, *The Physical Universe.* New York: McGraw–Hill, 1960.
10. K. Krauskopf, *Fundamentals of Physical Science.* New York: McGraw-Hill, 1959.
11. E. A. Fath, *Elements of Astronomy.* New York: McGraw-Hill, 1955.
12. F. T. Bonner and M. Phillips, *Principles of Physical Sciences.* Reading, Mass.: Addison-Wesley, 1957.
13. A. N. Strahler, *Physical Geography.* New York: Wiley, 1960.
14. *The Earth and Its Atmosphere,* D. R. Bates, ed. New York: Basic Books, 1957, pp. 32–33.
15. C. H. Deetz and O. S. Adams, *Elements of Map Projection,* 5th ed. Washington, D.C.: U.S. Government Printing Office, 1945.
16. A. N. Strahler, *The Earth Sciences.* New York: Harper & Row, 1963.
17. E. Raisz, *Principles of Cartography.* New York: McGraw-Hill, 1962.
18. W. L. Ramsey and R. A. Burckley, *Modern Earth Science.* New York: Holt, Rinehart and Winston, 1965.
19. F. L. Whipple, *Earth, Moon, and Planets.* Cambridge: Harvard University Press, 1963.
20. T. Page and L. W. Page, *Origin of the Solar System.* New York: Macmillan, 1965.
21. Earth Science Curriculum Project, *Investigating the Earth.* Boulder, Colo.: ESCP, American Geological Institute, 1965. (Text and Laboratory Manual.)

II. The Lithosphere

Chapter II is devoted to the solid earth, including the deep interior. For the title of this chapter we have adopted the term *lithosphere,* which means literally "the stone sphere." Many geophysicists use the term lithosphere in a more restricted sense, as referring only to the thin outer shell of the earth, called the crust. Indeed, because of its ambiguity, the term is not widely used at all. However, we shall continue to use it, but in the broadest sense, as referring to the entire earth, including its molten core but excluding the atmosphere and the oceans.

Among the scientific disciplines encountered in this chapter are those of geology, the study of the visible features of the earth's surface, seismology, the study of seismic waves in the earth and earthquakes, and geomagnetism.

Specific topics reviewed in the following sections include the geological history of the earth, the geological time scale as revealed by rock strata and by radioactive dating, and the geological processes responsible for the evolution of the earth's surface features. The physical structure of the earth's crust and the composition of the earth's deep interior as revealed by seismological evidence (earthquake waves) are examined next, followed by a discussion of the origin, distribution, and location of earthquakes. The final part of this chapter is concerned with the subject of geomagnetism. The magnetic field of the earth originates primarily in the earth's interior, and must be explained in terms of the earth's internal structure.

1. Geological history. *Historical note.* Since the eighteenth century, geologists have set themselves the task of describing the manner in which the earth has developed to its present state. Geological theory has oscillated between the extremes of *catastrophism,* the idea that the earth has changed through a series of unconnected cataclysmic events, and *uniformitarianism,* the idea that the earth has changed continuously and gradually by processes that are still going on. Modern geological theory is generally much closer to uniformitarianism than to catastrophism.

A. G. Werner (1749–1817), one of the founders of the science of geology, advanced the short-lived theory known as Neptunism, in which he proposed that all rocks originated as sediments from the sea. From a study of the rock strata near Freiberg, Germany, Werner concluded that the layering of sedimentary rock on top of granite was universal. He was thus led to believe that all rock strata were precipitated in sequence from a universal, primeval ocean from which the water subsequently had disappeared leaving behind the continents.

Another great figure of early geology was the biologist Georges Cuvier (1769–1832). One of the founders of the science of paleontology, Cuvier deduced from his study of fossils that different rock layers contained different forms of life, which were created separately and then destroyed by some cataclysm. By relating rock strata to the succession of biological events on earth, Cuvier pointed the way to the ultimate wedding of biological and geological evolution. But at the same time, his and Werner's view of separate and catastrophic events in geological history slowed the advancement of the science.

The first of the great opponents of the catastrophic theory was James Hutton (1726–1797), who viewed all geological events as parts of a continuous never-ending process. Hutton described a continuous cycle of rock formation from the sea, followed by wearing away by processes of erosion. He distinguished between volcanic and sedimentary rock, and laid the foundation for dating of rock layers (stratigraphy). Hutton's words, "the present is the key to the past," express his theory of uniform and continuous development. In emphasizing the dominant role of vulcanism as a geological process rather than sedimentation, Hutton became the leader of the so-called Plutonist school of geological thought. The controversy between the Neptunians and the Plutonians, which raged for about fifty years, ended with the rejection of Werner's hypothesis.

Geology owes much to the tireless engineer, William (Strata) Smith (1769–1838), who mapped the rock layers of England, showed the similarity of the rock sequences in different locations, even where separated by valleys or water, and, as one of the great stratigraphers, contributed as much as any man to the biography of the earth.

In the nineteenth century, catastrophism was further weakened by the work of Charles Lyell (1797–1875), who sought to prove that the processes which are now in operation on the earth are capable of accounting for all the geological events of the past, if we consider them to have operated over a very long time. When Darwin's *Origin of Species* appeared in 1859, Lyell was able to accept the theory of continuous evolution for geology as well as biology, and to see the unity of evolutionary development in life and in the earth.

Gaps and discontinuities in the stratigraphic record, however, could be explained only by assuming that there had indeed been periods in the past when the intensities of geological processes were different from what they are now. Thus the earth has experienced intermittent periods of mountain building and ice ages which, superimposed on the gradual development of the earth, help to explain its present state.

The geological time scale. The dates of geological history have been determined in several ways. The relative ages of rock strata can be ob-

TABLE 4
GEOLOGICAL HISTORY

Time of beginning, millions of years	Era	Period	Epoch	Biological events
0.01 (10,000 years)	Cenozoic ("new life")	Quaternary	Recent	
1			Pleistocene	First men. Last ice age
13	Age of mammals	Tertiary	Pliocene	
25			Miocene	
36			Oligocene	Mammals develop
58			Eocene	
63			Paleocene	
ROCKY MOUNTAIN REVOLUTION				
135	Mesozoic ("middle life")	Cretaceous		Last dinosaurs
181		Jurassic		First mammals
230	Age of reptiles	Triassic		First dinosaurs
APPALACHIAN REVOLUTION				
280	Paleozoic ("ancient life")	Permian		Salt deposits
345		Carboniferous		Coal. First reptiles
405		Devonian		First amphibians
425		Silurian		Land animals and plants
500		Ordovician		Fish
600 (?)		Cambrian		Marine invertebrates
KILLARNEY REVOLUTION				
More than 3400	Precambrian			Fossil algae

tained by studying the sequences of the layers. Further information is derived from measurements of the thicknesses of the layers and from a knowledge of the average rate of deposition of the sediments and average rates of erosion. Tree rings and clay varves (seasonal layers of glacial sediment found in lakes) provide additional data on the ages of rocks. Rock stratigraphy is supplemented by paleontology, the study of fossil remains, which adds to our knowledge of the ages of rock strata and the dates of geological and biological events. Certain "index fossils" clearly date the rock strata in which they are found. Finally, to determine the absolute time scale of geologic events, we turn to the measurements of radioactivity in rocks and fossils. Work on radioactivity now in progress in laboratories may result in further refinements of the geological time scale.

The geological history of the earth is summarized in Table 4. The time data listed in the table are based on recent radioactivity measurements. The geological eras are named in accordance with paleontological evidence, i.e., on the basis of the kind of life that did or did not make its appearance at the time the rock strata were laid. The geological periods correspond to the rock strata, and in some cases the names of the periods refer to the locality where a particular stratum, completing a geological sequence, was discovered.

It is convenient to divide the geological record into four eras separated by three periods of mountain building, called revolutions. However, many geologists are careful to point out that these "revolutions" were neither abrupt nor simultaneous over the whole earth, and therefore do not really separate the eras except in the broadest sense.

The oldest rocks on earth formed more than 3.4 billion years ago in the *Precambrian era*. Relatively little is known about the geological and biological processes during this first era in the history of the earth. Following the solidification of the earth into rock in early Precambrian time, conditions favorable for life evolved. There is now evidence that primitive forms of life appeared on earth in the Precambrian era. Fossil algae as old as 2.8 to 3 billion years have been found in rocks of the Precambrian era. Precambrian rocks are found at the surface over broad areas of the continents, e.g., over Canada and the United States.

The *Paleozoic* (Gr. *palaios,* ancient, plus *zoe,* life) *era* began about 600 million years ago following the Killarney Revolution (in North America), the mountain-building period that produced the Adirondacks and the now eroded mountains near Lake Superior. In this mild era, life appeared on earth in abundance. Marine fossils have been found in the oldest Paleozoic rocks, the Cambrian, named after Cambria, the ancient name for Wales, where these rocks were first found. In the later Paleozoic periods—Ordovician (named after a Welsh tribe), Silurian (another Welsh tribe), Devonian (after Devonshire, England), Carboniferous (including the Mississippian and Pennsylvanian periods, when great deposits of coal-forming vegetation were laid down), and Permian (from Perm, in eastern Europe, where these strata occur)—many new forms of life appeared, including fish, land animals, amphibians, and reptiles.

The Paleozoic era ended with the period of upheaval and mountain building known as the Appalachian Revolution, when the Appalachian Mountains of North America formed. About 230 million years ago, the *Mesozoic* (Gr., *meso,* middle, plus life) *era,* which is known as the age of reptiles, began. In the Triassic period of this era the great dinosaurs appeared, and by the end of the era (about 100 million years ago) they had perished. In the Jurassic period (named for the Jura Alps of France) the first mammals appeared, and in the Cretaceous period (a term referring to chalky deposits) flowers formed.

Following the Rocky Mountain Revolution (about 100 million years ago) our current era, known as the *Cenozoic* (Gr. *cene,* new, plus life) began. This era, also known as the age of mammals, is divided into two periods: the Tertiary and the Quaternary (third and fourth quarter, from an older classification system). The five epochs of the Tertiary period, extending from Paleocene ("oldest new") about 63 million years ago to Pliocene ("more new") which ended about one million years ago, saw the evolution of higher forms of life. Man first appeared in the Pleistocene ("most new") epoch of the Quaternary period. This epoch was marked by the last great ice age, which ended about 10,000 years ago, at the beginning of the epoch we call *Recent.* The Pleistocene epoch is sometimes referred to as the great ice age.

Geological processes. We may classify the processes responsible for the development of the earth's surface features into three groups: building processes, leveling processes, and transformation processes.

Among the building processes, i.e., those which elevate the land masses, are *vulcanism, sedimentation,* and some kinds of *diastrophism.*

Vulcanism is the movement of molten rock, called *magma,* beneath the earth, and its extrusion through fissures onto the surface of the earth, where it pours out of volcanoes as lava. Whether explosive or quiet, all active volcanoes are in the process of building mountains. Volcanic rocks are referred to as *igneous rocks,* i.e. rocks of fiery origin. Igneous rocks that have formed within the earth by slow cooling are called *intrusive.* Those which have formed by rapid cooling on the earth's surface are referred to as *extrusive.* The former are coarse grained (e.g., granite), and the latter are fine grained (e.g., basalt), or even amorphous and glassy (e.g., obsidian). Gas in the volcanic lava may make it light and full of holes (e.g., pumice). Igneous rocks containing much silica (silicon dioxide, the mineral quartz) are light colored (e.g., granite, rhyolite), while those with little silica and much metallic oxide are dark colored (e.g., gabbro and basalt).

Sedimentation is the process of rock formation by deposition of rock fragments, sand, clay, or organic materials. The sediments are products of erosion or remains of life which have been transported by wind, water, and glaciers. Prolonged accumulation of the sediments leads to formation of deep rock strata. In time the weight of a column of sediment compresses the older, underlying sediments into rock. Among the sedimentary rocks are the conglomerates (cemented gravel), sandstone (composed of sand-size particles), shale (consolidated mud or clay), and limestone (composed of calcite, i.e., calcium carbonate, precipitated from water or derived from shell or skeletal fragments, and including chalk). Fossils, the preserved remains of ancient life, are found in many sedimentray rocks.

Diastrophism (Gr. *diastrophe,* distortion) refers to processes which deform the earth's crust. The internal earth-shaping forces which produce these deformations are called *tectonic* forces. Three basic diastrophic processes are fracturing (the development of cracks or faults in the rock), folding, and warping. Folding results in wavelike beds of rock. Rock layers humped up in crests are called anticlines, while those dipping down in troughs are called synclines. At times great areas of the continents adjacent to the oceans have subsided, permitting the sea to cover the land. Sediments then accumulated in these shallow inland seas about as rapidly as the land subsided. Such regions are known as geosynclines. Later upwarping of the geosynclinal regions have thrust these marine sediments well above sea level, as in the Appalachian and Rocky Mountains of North America.

Faults may result from unequal loading of sediments on underlying rock, or from upward pressures on the crustal material caused by internal forces in the earth. The stretching of the nonplastic rock material produces the cracks known as faults. In a *normal* fault one rock layer, represented by the hanging wall, has moved downward relative to the adjacent rock, represented by the foot wall. In a *reverse* (or *thrust*) fault the hanging wall has moved up relative to the foot wall. A *strike-slip* fault is a nearly vertical fault along which the rock layers have been displaced horizontally.

The leveling processes on the earth are those of erosion. Erosion may be caused by weathering, by wind, by streams of water, by wave action on shores, and by movements of ice. Weathering is the wearing away of rocks by chemical action of air (e.g., oxidation), by rain, which has both chemical* and mechanical effects, and the breaking of rocks caused by the expansion of water upon freezing in fissures. Wind produces two kinds of erosion, known as *deflation,* the removal of loose particles, and *abrasion,* the wearing away of rocks by the sand and dust carried by the wind.

---

* For example, carbon dioxide dissolves in water to produce carbonic acid, which in turn dissolves limestone.

Streams of water, probably the most important agents of erosion, cut valleys through sediments and rocks. Young valleys are V-shaped but as the river grows older it meanders and the valley grows broad, until eventually a wide flood plain has formed. Waves beating upon the shores of oceans and smaller water bodies carry away loose particles and also wear down the rocks. Not the least of the agents of erosion are moving glaciers, which have produced and continue to produce major changes in land forms by scouring the underlying rock and transporting debris (rocks and soil) known as glacial *moraine* or *till*.

An important rock-forming process in the crust is metamorphism, the transformation of rocks by heat or pressure below the earth's surface. Both igneous and sedimentary rocks, as well as other metamorphic rocks, may be metamorphosed. Examples of metamorphism are the transformation of shale into slate, the transformation of certain igneous, sedimentary, and metamorphic rocks into schist, and the transformation of igneous and sedimentary rocks into gneiss, a coarse-grained, banded rock, rich in mica, which looks somewhat like granite. The metamorphosis of limestone produces marble, and that of sandstone, quartzite.

Mountain building and the origin of the continents. All the young mountains and islands of the earth lie along two long chains or arcs which represent two great fractures in the continental structure. These are the *orogenetic* (Gr. *oros,* mountain, plus *genesis,* creation) belts of the earth along which mountain building has been going on most recently. Both regions are characterized by two features: they contain the majority of the world's active volcanoes (almost 500 of them), and they are regions of maximum earthquake activity. One of these orogenetic belts is the Eurasian-Melanesian belt, extending eastward through the Mediterranean, across the Alps and Himalayan Mountains, to Indonesia, the Solomon Islands, and New Zealand. The other, known as the East Asian-Cordilleran belt, extends around the coasts of the Pacific Ocean, through Japan, Alaska, and the Pacific Coasts of North and South America.

The orogenetic belts represent the regions of most recent mountain building and continental development. In the geologica past there have been several similar revolutions. Geologists have sought for years to explain how the surface features of the earth, especially the continents and mountains, have formed. We know that the primary mechanisms for mountain building are vulcanism and the thrusting upward of the earth's crust by forces in the earth. But the details of mountain and continental development are still not well understood.

One theory of the development of the continents is known as the *theory of continental drift.* According to this theory, large continents broke apart into smaller land masses which drifted away from one another, leaving the oceans between them. This interesting theory is still a subject of considerable controversy among geologists. Recent studies of the ancient magnetic fields of the earth (see page 64) appear to lend much support to the theory that such relative motions of the earth's crust may indeed have occurred in the past.

Another theory of continental development, known as the *convection theory,* holds that the material of the earth's mantle, the layer below the crust, flows in currents, like a fluid. Piling up of the earth's mass by converging currents would account for the development of both mountains (by lifting) and fractures in the earth.

A third theory of continental development is the *contraction theory.* According to this theory, the earth below a depth of about 700 kilometers (400 miles) is no longer cooling or shrinking. Moreover, the upper layer of the earth, above 100 kilometers (60 miles), is also in a state of equilibrium. We have, then, an earth in which cooling and contraction during most of geological time have been confined to a layer sandwiched between depths of 100 and 700 kilometers. The contraction of such a shell, constrained between two adjacent unshrinking shells, results in the tectonic forces which have molded the earth. As the cooling shell of the mantle contracts, tension produces fractures in the layer along which both the mantle and crustal material slip. It is believed that the great continental fracture system of the earth—the orogenetic arcs—may have been produced by such a contraction process. The contraction theory does not explain the building of mountains, except perhaps indirectly. Tension in the deep cooling shell is accompanied by compression of the outer layer of the earth, resulting in volcanic and diastrophic effects in the crust which can produce mountains, islands, and continents.

Still another theory of the earth's development is the *expansion theory,* according to which the earth is continually expanding. The expansion process would explain the great rift (crack) that has recently been discovered running through the ocean floor around the earth. However, the idea that the earth is expanding is rejected by most geophysicists.*

2. The earth's crust. *Layers of the earth.* The earth is composed of several layers, like an onion. The existence of these layers and the depths of the boundaries between them have been deduced from the study of earthquake waves passing through the earth. The deepest layer, at the center of the earth, is the solid *inner core.* Surrounding this layer is a molten *outer core.* Above the core rests the layer known as the *mantle.* And resting on the mantle is the thin outer shell of the earth called the *crust.*

The boundary between the crust and the mantle was discovered by the geologist Mohorovicic in 1909. Mohorovicic observed that seismic waves passing through the earth were bent in a manner that indicated an abrupt change, or discontinuity, in the earth's composition and density with depth

---

* See references 1 through 11.

at about 35 kilometers below the earth's surface. This boundary between the crust and mantle was later named the Mohorovicic discontinuity, and is usually referred to as the "Moho."

There is considerable interest today in the Moho. The mantle, which lies below the Moho, has never been seen by man. It is believed that much could be learned about the formation of the earth from an examination of the material making up the mantle. Therefore, it has been proposed that a hole be drilled through the crust and Moho, into the mantle. Plans have been made to drill this hole (known as the "Mohole") under the sea, where the crust is thin.

Depth of the crust. The depth of the crust is greater under the continents than under the oceans, as shown schematically in Fig. 27. The continents are composed of deep blocks of granitic rock, probably resting on a basement of basaltic rock. The depth of the Moho under the continents is about 35 kilometers (about 20 miles). Under the oceans, the granite blocks are absent. The ocean-bottom sediments rest directly on a basement of basalt, which is only about five kilometers (three miles) thick. The advantage of drilling down to the Moho from the ocean floor rather than from the continent is clear from the figure.

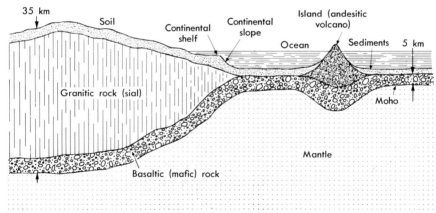

Fig. 27. The crust of the continent and ocean.

Composition of the crust. The continental blocks are composed of relatively light, siliceous rock, such as granite, andesite, and various grades of metamorphic rock, known under the general name *sial* (aluminum silicate). The heavier basaltic rocks which constitute the crust under the oceans contain higher proportions of metallic oxides. Relatively rich in iron and magnesium, the minerals composing these rocks are referred to as ferromagnesian minerals. The basaltic rocks are referred to collectively as *mafic*.

On the basement of igneous and metamorphic rocks which constitute the stable continental shields where no volcanic action or mountain building is now taking place there rests a veneer of soil (unconsolidated sediments) and sedimentary rocks. The sediments encountered under the oceans are found to be of three kinds. Near the coasts, on the continental shelf and continental slope (see Fig. 27), the sediments are mud and clay that have been washed off the land by waves; these are called *terrigenous* sediments. The sediments found in deep water are called *pelagic* sediments and are of two kinds: organic material called *oozes,* and inorganic material, called *red clays,* which may be composed largely of meteoric dust. Oceanic sediments accumulate at a rate of approximately one meter (about 40 inches) every million years. The layers of sediments under the Atlantic are thicker (recent measurements in the Atlantic Ocean suggest that the sediments there vary in thickness from 500 to 2500 meters) than those under the Pacific Ocean (less than 400 meters), where they have apparently been covered by lava from recently active volcanoes.

The crust beneath the oceans. More than 70 percent of the earth's surface is covered by oceans. The crust beneath the oceans is as irregular and rugged as that of the continents. Measurements of ocean depths have been carried out on oceanographic expeditions for many years, first with sounding lines (weighted wires) and later with echo sounders, which use the reflection of sound waves to measure the depth of the ocean bottom. These observations show not only that the average depth of the oceans is about 4 kilometers (about 12,000 feet), but that the oceans are crisscrossed with mountains and valleys.

Adjacent to the coastline, the shallow-water terrigenous sediments form the *continental shelf,* whose depth below sea level is no greater than about 200 meters (600 feet). (Depth in the ocean is also expressed in fathoms, the fathoms being 6 feet. Thus the maximum depth of the continental shelf is about 100 fathoms.) The bottom then falls abruptly with distance from the coast along the *continental slope.* Both the slope and shelf are indented with canyons and valleys, for example, the submarine Hudson Valley.

Around the edges of the deep oceans are found broad *basins* containing the mediterranean seas of the earth. In the deep oceans, the deeply submerged basins are called *deeps.* Running through the oceans are deep valleys called *troughs,* and cutting through some of these troughs are the deepest portions of the oceans, the steep-walled submarine canyons called *trenches.* The deepest trenches are found in the western Pacific Ocean in the vicinity of Guam, the Philippines, and the Marianas Islands—along the East Asian orogenetic arc. The Navy bathyscaph "Trieste" recently descended to a depth of almost 12 kilometers (about 7 miles) in the Marianas Trench.

Elevations of the ocean bottom vary from small islands to great submarine mountains, such as the Mid-Atlantic Ridge which divides the Atlantic along a north-south line running through the Azores. In shallow water are encountered broad submerged plateaus called *banks* or *shoals,*

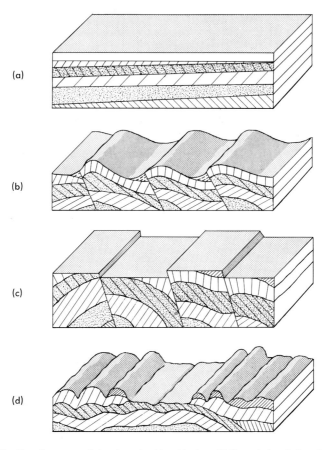

Fig. 28. Development of the Appalachian-Hudson Valley region (after Strahler).

the latter term being used for banks whose closeness to the surface endangers navigation. Shoals whose rock is exposed at or above the water surface are called *reefs*. The ridges separating the submarine basins are referred to as *sills*. Submerged volcanic hills, too low to be islands, are called *seamounts* or guyots. Frequently the unconsolidated marine sediments are washed off the seamounts, exposing the underlying bedrock of the oceanic crust.

An important discovery regarding the oceanic crust has been that of the great *rift* in the ocean bottom, a crack in the ocean floor which appears to be worldwide, and which may possibly be continuous through the continents.

Examples of crustal development. The morphology of the earth's crust results from the action of many geological processes. The development of the Appalachian-Hudson Valley region, shown in highly schematic fashion in Fig. 28, illustrates some of these processes.

(a) In very early Paleozoic times both the Cordilleran (Rocky Mountain) and Appalachian regions were occupied by shallow inland seas which flooded broad geosynclines and laid down thousands of feet of sediment, which later metamorphosed into layers of shale, slate and schist, and finally gneiss.

(b) At the end of the Paleozoic era (the Appalachian revolution), mountain-building processes produced folds, thrust faults, and general upwarping (lifting) of the crust above sea level.

(c) Erosion in the Mesozoic era formed a peneplain ("almost-a-plain"). Block faulting led to the appearance of fault scarps (cliffs), but further erosion again leveled the surface and a new peneplain formed.

(d) Upwarping followed by stream erosion produced the present ridge (hard rock) and valley (soft rock) system. In the Pleistocene epoch the terrain was further modified by glacial action.

The gradation of the earth by glaciers is another major factor influencing the shape of the crust. Ice ages have occurred at intervals of about 250 million years. The latest of the ice ages, the Quaternary or Pleistocene Ice Age, began about one million years ago. During the Quaternary there have been four glacial (cold) periods and three or four interglacial (warm) periods on earth. We are now in either an interglacial or a postglacial period. The last glacial period, the Wisconsin, ended about 11,000 years ago.

During an ice age the perennial ice extends to low latitudes, in the form of continental ice sheets, mountain (alpine) glaciers, or sea ice. The first two alter the underlying terrain as they advance.

Alpine glaciers scour out U-shaped valleys, as contrasted with the V-shaped valleys cut by streams. A fjord is an example of a glacial valley that has been filled by the sea. Some other features of alpine topography created by glaciers are cirques (bowls), arêtes (the ridges between two cirques), and horns (the jagged points between three or more cirques).

As the glaciers move across the continents they polish, grind, and cut grooves or striations (*striae*) in the underlying rock, partly by scraping against it with transported debris. Hills or knobs are shaped by the moving glacier. On the lee side of the knob, the land slopes steeply, while on the opposite side, called the "stoss" side, the land slopes gently. Glacial lakes, e.g., New York's Finger Lakes, are among the products of glacial advance.

The rock and soil fragments transported and deposited by the glacier are called glacial till or moraine. Surface, or *lateral,* moraine falls onto the side of the glacier from the valley walls. *Medial* moraine lies along the center of a valley glacier formed by the confluence of two or more valley glaciers. *Terminal* moraine is carried along at the tongue (forward edge) of the glacier, and left behind when the glacier retreats. The streamlined hills of glacial till shaped by the moving ice are called *drumlins*. The elongated ridges of till deposited by glacial streams are known as *eskers*. Both drumlins and eskers are usually relatively treeless compared with the surrounding terrain.*

3. The interior of the earth. Most of our information about the earth's interior is derived from the study of seismic (earthquake) waves. Thus the subject of seismology is inseparable from that of the earth's interior. We first discuss seismology with a view to the information it gives us about the internal structure of the earth. This approach will be followed by a treatment of seismology as the study of the earthquakes themselves.†

Historical notes. Scientific seismology began in 1760 with the recognition by Mitchell that earthquakes send out *seismic waves* which travel through the earth. The first useful seismometer for measuring these earthquake waves was invented in 1855 by Palmieri, who also coined the word seismograph (Gr. *seismos,* earthquake). However, the first true seismograph (recording seismometer) was built by Milne in 1892.

About 1890, Wiechert proposed the hypothesis that the core of the earth is in a fluid (molten) state. The molten-core hypothesis was proved with seismological evidence by Oldham in 1906. In 1909, Mohorovicic discovered the discontinuity between the crust and the mantle. Five years later, in 1914, Gutenberg determined the thickness of the mantle and the radius of the earth's core, and showed that the earth exhibited other discontinuities in addition to the Moho. In 1936, Lehmann presented the theory, based on seismological evidence, that the entire core may not be fluid, and that there may be a solid inner core at the earth's center.

Seismic waves. An earthquake transmits energy through the earth in the form of waves. It is these waves which are felt as earth tremors, even at a considerable distance from the source. The movements of the earth's crust associated with seismic waves are measured with a seismograph. A seismograph consists of a heavy mass supported by a spring, which shakes (up and down or sideways, depending on whether we wish to measure the vertical or horizontal tremors) relative to the recording apparatus when the earth shakes. The relative motion of the mass is due to its inertia. The mass is actually stationary, and the earth is moving relative to it. A pendulum may also be used as a seismograph. The pendulum hangs vertically when the earth is still. When the earth shakes, the pendulum mass remains stationary due to its inertia, but the earth, the point of suspension, the case of the instrument, and the recording chart all move relative to the inert pendulum mass.

---

* See references 1 through 11.
† Section 4.

CHAP. 2] THE INTERIOR OF THE EARTH 55

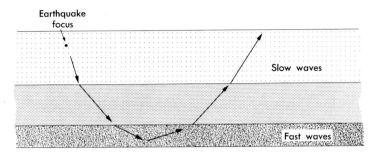

Fig. 29. Refraction of a seismic wave, showing the bending of the seismic ray as it passes through rock layers of different wave speed.

There are three types of seismic waves:
1. Primary or preliminary waves, called *P*-waves. These are the fastest waves. They travel through the earth at a speed of 5 to 13 kilometers per second (3 to 5 miles per second), and therefore are the first waves to arrive at the seismograph from the earthquake. They are of small amplitude and short period (0.5 to 5 seconds between waves). *P*-waves are similar to sound waves, i.e., they are *longitudinal* waves which travel by compression and rarefaction of the earth (in the manner of the "slinky" coiled-spring toy).

Like sound waves, *P*-waves can be bent or *refracted*. Refraction occurs when the wave passes from one kind of material into another where the speed of the wave is different. For example, when the wave passes from one rock layer into a less compressible layer, the speed of the wave increases and the wave bends as shown in Fig. 29. From the observed refraction of the waves, seismologists can determine the change in the physical properties of the earth with depth. At discontinuities in the earth, such as the Moho, seismic waves may be refracted to the point of being totally reflected. (See Fig. 29.)

The speed of *P*-waves depends on the density of the rock, its rigidity, and its compressibility. In solid rocks, the wave speed increases with increasing depth due to the decrease of compressibility of rock with increasing depth. As a result, the waves bend as shown in Fig. 29.* (When the wave passes from a "slow-speed" layer to a denser "high-speed" layer, it bends away from the vertical.) However, when the *P*-wave enters a layer of *fluid,* its speed abruptly decreases. Thus, as the *P*-wave penetrates down into the molten core of the earth, it slows down, and is bent downward. This phenomenon gives rise to the existence of a *shadow zone* for *P*-waves

---

* The refraction of the seismic *ray* is shown in Fig. 29. The ray is the wave path, and is perpendicular to the wave front. It represents the direction along which the wave energy travels.

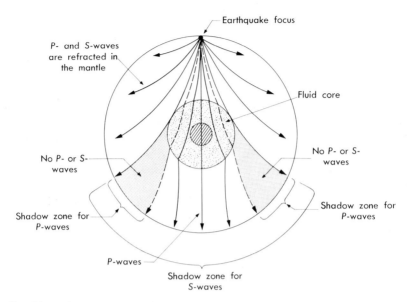

Fig. 30. Refraction of seismic P-waves in the mantle and core, showing shadow zone for P-waves. The larger shadow zone for S-waves is due to the absorption of S-waves by the earth's fluid core.

(see Fig. 30), which was one of the facts leading to the discovery that the earth's core is molten.

2. Secondary waves, called S-waves. These waves travel more slowly from the earthquake than do the P-waves (about two-thirds as fast). They also are of larger amplitude and have longer periods than P-waves.

S-waves differ from P-waves in another important respect. They are *transverse* waves. This means that the oscillations are not back and forth along the direction of the wave propagation, as in a longitudinal-compression wave. Instead, the oscillations are lateral, perpendicular to the direction in which the wave and its energy travel. S-waves are also called *shear* waves, because the material through which the wave travels experiences a shearing deformation instead of compression.*

An important feature of transverse waves is their inability to travel through fluids because a fluid cannot be "sheared." Thus S-waves which enter the molten core of the earth are absorbed there, and do not penetrate the fluid core. The observation that S-waves emanating from earthquakes do not appear on the opposite side of the earth gave strong support

---

* To visualize the shearing process, place a book facedown on a table, press down on the cover with your hand, and, holding the book firmly against the table, slide the top cover back and forth relative to the bottom cover. The resulting motion is called a shearing motion.

to the theory that the earth has a molten (fluid) core. The S-wave shadow zone is illustrated in Fig. 30.

3. Surface waves, also known as Rayleigh and Love waves. These are very slow, long-period waves of large amplitude, which travel through the earth's crust, like water waves, but do not penetrate into the interior.

Layers of the earth. The existence of the molten core of the earth was deduced from the existence of the shadow zones for P- and S-waves, illustrated in Fig. 30. The depth of the boundary between mantle and core was calculated by reconstructing the refracted P- and S-waves from earthquakes, and by measuring the width of the shadow zones.

On occasion it has been observed that some P-waves appear in what should be the P-wave shadow zone. Such ray paths are possible only if the molten core contains a solid inner core in which the direction of refraction is again reversed. Thus it has been proposed that the core consists of two shells: a fluid outer core, and a solid inner core.

TABLE 5

THE LAYERS OF THE EARTH, AND THE VARIATION OF DENSITY WITH DEPTH

Depth below surface		Density (relative to that of water)	Layer
km	mi		
0	0	2.5	Crust
33	20	3.3	Moho
2900	1800	5.7	Bottom of mantle
2900	1800	9.4	Top of outer core
5100	3200	14.2	Bottom of outer core
5100	3200	16.8	Top of inner core
6370	3960	17.2	Center of earth

The density of the earth increases with increasing depth toward the center of the earth. Table 5 shows the variation of density with depth as one penetrates through the several layers that constitute the earth. (Note that the density of water is one gram per cubic centimeter.) We see that the density jumps discontinuously as we cross the boundaries between layers. Note also that the depth of the mantle is less than half the radius of the earth, the remainder of the earth being occupied by the inner and outer cores. The thicknesses of the layers are: crust, 5 to 30 kilometers; mantle, 2900 kilometers; outer (liquid) core, 2200 kilometers; inner (solid) core, 1270 kilometers.

Temperature of the earth. Measurements of temperature in deep mine shafts show that the temperature increases toward the center of the earth. In the upper crust the temperature probably rises at a rate of about 20 degrees centigrade (°C) for every kilometer of depth (or about 60 degrees fahrenheit per mile). However, the temperature cannot rise at this rate all the way into the interior, for if it did, the mantle would not be solid.

It is estimated that the earth's temperature may rise to perhaps 400 to 700°C at the Moho, to about 2500°C at the bottom of the mantle, and to perhaps 3000 to 5000°C at the center. There exists within the earth a vast store of energy in the form of heat that may someday be tapped by man for his own use.*

4. Earthquakes. It has been estimated that about one million earthquakes occur each year. Most of them remain undetected because they occur in uninhabited regions (e.g., the oceans) or are too weak to be recorded. About 40,000 earthquakes a year are detectable without instruments. Of these about 100 a year have destructive effects, and about once a year there occurs a really "great earthquake." Almost all earthquakes occur in the orogenetic belts associated with the great continental fracture system.

Seismographs. Seismographs (Fig. 31) are designed to record the seismic waves emanating from the source of the earthquake. The seismograph records the *P*-, *S*-, and surface waves (Fig. 32) from the earth-

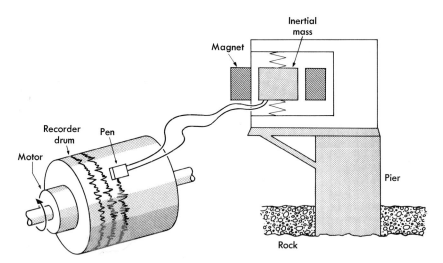

Fig. 31. Schematic drawing of vertical component seismograph (electromagnetic type). The recording drum rotates continuously and advances from left to right.

---

* See reference 12 and also references 1 through 11.

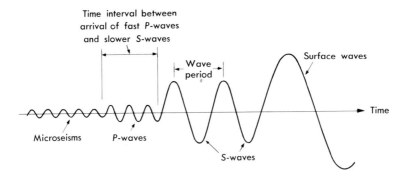

Fig. 32. Seismic waves as recorded by a seismograph.

quakes as well as the small background oscillations, called microseisms, produced by wind, ocean waves, and other causes. In addition to the *initial P-* and *S-*waves (or *phases*), *reflected* waves (called *PP, PPP, SS, SSS,* etc.) also arrive at the seismograph and are recorded. These waves are reflected from the surface of the earth or from internal discontinuities (often more than once), and therefore arrive at the seismograph after the initial phase. This, together with the series of shocks emanating from the focus of the earthquake, accounts for the complicated appearance of the earthquake seismogram.

To record both the vertical and the horizontal shaking of the earth, it is necessary to have two seismographs which differ in the way the inertia masses of the instrument are mounted. A seismograph must be mounted on a pier embedded in the bedrock (see Fig. 31). The inertial mass is suspended from a support attached to the pier, so that the pier moves relative to the mass when the earth trembles. To complete the instrument, a device is needed for amplifying and recording the relative oscillations of the mass. A damping device is also employed to filter out unwanted oscillations. In one type of seismograph, the inertial mass moves between the poles of a magnet. A coil of conducting wire is wound about the mass so that it behaves like an electric generator. The electric current generated in the coil increases with the rapidity of its motion. Thus the seismic waves can be recorded electrically.

In seismographs used for recording natural earthquakes, the instrument is damped in such a way that waves with a period close to a few seconds are magnified relative to all others. However, in exploration seismographs, which record the waves from man-made explosions, the damping is designed to magnify much shorter period waves.

Seismologists express the intensity of an earthquake on a scale, known as the modified Mercalli scale, which is based on the destructive effects of the earthquake. The intensities of earthquakes on the Mercalli scale range from I ("felt only by a few people; no visible effect") to XII ("panic; total destruction; waves seen on the ground; objects thrown in the air"). The San Francisco earthquake of 1906, which killed 390 people, is estimated to have been of intensity XI. The most destructive earthquake of all time, the Lisbon earthquake of 1755 which killed 55,000 people and shook one million square miles of Europe and Africa, was one of the scale-XII earthquakes. The most violent earthquake known to man occurred in Assam, India, in 1897. The Assam earthquake occurred in a relatively sparsely inhabited area, so that, despite its fantastic effects on the earth, only 1000 to 1500 people died in this catastrophe. In more recent times, disastrous earthquakes have continued to occur along the Mediterranean orogenetic arc, the Cordilleran arc on the Pacific coast of South America, and elsewhere. On February 29, 1960, the city of Agadir, Morocco, was destroyed by an earthquake that left almost 12,000 dead. Less than three months later a series of earthquakes and associated tidal waves, volcanic eruptions and landslides killed an estimated 5000 victims in central Chile. As yet there is no way of predicting where or when these disasters will occur.

From the destructive effect of an earthquake one cannot determine its absolute *magnitude*. (The distance of an earthquake from an inhabited region, for example, is at least as important as its magnitude, as far as destruction is concerned.) To express earthquake magnitude, seismologists use a scale, the Richter scale, which takes into account the distance of the earthquake from the seismological station. The magnitude of the earthquake on the Richter scale is determined from the seismograms produced by the earthquake, rather than from the destruction wrought.

Location of earthquakes. The destructive effects of earthquakes are produced by the surface waves close to the source of the shock. The location of the earthquake, however, is determined by studying the propagation of the *P*- and *S*-waves.

The *focus* of an earthquake is the point *in* the earth where the earthquake originated. The *epicenter* of the earthquake is the point *on* the earth's surface directly above the focus. To locate the epicenter of an earthquake, three seismological stations are needed. If the distance of the epicenter from each station can be determined, a circle with that radius can be drawn about each station. The point at which the three circles intersect is the epicenter of the earthquake.

The distance of the earthquake epicenter from a seismograph is determined from the time interval between the arrival of the first *P*-waves and that of the first *S*-wave (see Fig. 32). The velocities of the two types of wave are known. The time interval between the arrivals of the waves, therefore, depends only on the distance the wave has traveled from the focus. Thus, from the time interval, which is measured on the seismogram, the distance to the focus (and epicenter) can be determined at each seismological station.

The cause of earthquakes. Seismologists now believe that all significant earthquakes are produced by tectonic forces in the earth. It is generally agreed that other possible causes of earthquakes (meteorites, rock slides, cave-ins, and volcanoes) are incapable of producing any major earthquakes.

Earthquakes originate in the slipping of rock strata along fault planes in the crust following a period of stress and deformation of the rock. The rupture or *faulting* of the rocks is the result of tectonic forces within the earth. One of the largest and best known faults is the San Andreas fault, which extends the length of California and is responsible for the earthquake activity in that region.

It is generally accepted that the principal cause of earthquakes is the phenomenon of *elastic rebound*. Horizontal tectonic forces deform the rock through shearing motion until the elastic limit of the rock is exceeded and the earth snaps back (rebounds) to produce a fault, and possibly an earthquake.

Earthquakes are classified, according to their depth of focus, as *normal* (or *shallow*) *focus* (0 to 70 kilometers deep), *intermediate focus* (70 to 300 kilometers), or *deep focus* (greater than 300 kilometers). Normal-focus earthquakes, which occur usually in the crust, are about five times as frequent as intermediate-focus earthquakes, and about twenty times as frequent as deep-focus earthquakes. The latter have never been recorded at depths greater than 700 kilometers (450 miles).*

5. Geomagnetism. Compass needles were used for navigation in Asia at least as early as the eleventh century. However, the reason for the behavior of the compass needle was first explained by William Gilbert (1540–1603), physician to Queen Elizabeth, in his treatise *De Magnete*, published in 1600. Gilbert stated that the earth is itself a magnet, with a north and a south magnetic pole. The north-pointing pole of the compass needle is attracted by the north magnetic pole of the earth, whereas the south-pointing pole of the compass is attracted to the earth's south pole.

Gilbert believed that there was a permanent magnet within the earth (as in a demonstration model of a magnetic globe). It is now believed that the interior of the earth is too hot for permanently magnetized material to exist there. The explanation now generally accepted for the earth's magnetism is that there are electric currents within the core (due to its fluidity and motion), which produce a magnetic field, as in an electromagnet. This *dynamo theory* has been developed only during the last 15 years.

Some aspects of magnetism. A magnet, whether it be a magnetic mineral, such as magnetite, an iron bar in which magnetism has been induced,

---

* See references 13 and 14 and also references 1 through 12.

or an electromagnet, has two poles, designated north and south. A pair of magnetic poles located at a point is called a dipole. A north pole placed near the north pole of the magnet will be repelled, whereas a south pole placed near the north pole of the magnet will be attracted to it. The end of the compass needle which points toward the magnetic north pole of the earth is thus a "south" pole. To avoid confusion, we refer to this end of the compass needle as the *north-seeking* pole of the compass, and to the other end as the *south-seeking* pole.

The force of attraction or repulsion between two magnetic poles varies inversely as the square of the distance between them. Every magnet is surrounded by a magnetic field. The intensity of the field at any point is measured by the force which it would exert on a unit magnetic pole placed there. The intensity, or strength, of the magnetic field is expressed in terms of a unit called the *gauss* (after K. F. Gauss, nineteenth-century mathematician and geophysicist). The intensity of the earth's magnetic field is less than one gauss.

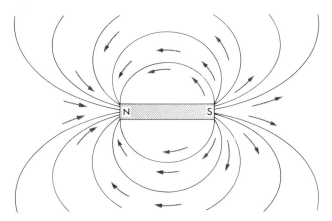

Fig. 33. Magnetic line of force. (The arrowheads represent the north-seeking pole of a compass needle.)

The magnetic field surrounding a magnet may be represented by *lines of force,* as illustrated in Fig. 33. The lines of force surrounding a bar magnet may be demonstrated by sprinkling iron filings on a sheet of glass or paper near the magnet. By a process known as induction, the slivers of iron become magnetized by the magnetic field. Each iron splinter then acts like a compass needle, and orients itself parallel to the magnet lines of force. The magnetic force, or field intensity, at any point is directed parallel to the lines of force. Note how the direction of the magnetic field varies around the magnet.

In 1820, Oersted showed that a magnetic field is produced by an electric current flowing through a wire. If the current flows through a straight wire, the magnetic field forms rings around the wire. If the current flows through a circular loop (coil) of wire, the magnetic lines of force ringing the wire are perpendicular to the coil at the center of the loop. The magnetic field of the earth may be measured by comparing it with the field produced by an electromagnet consisting of several coils of wire through which flows a known current. This is one type of magnetometer. In other types of magnetometers, the magnetic field strength of the earth is determined by measuring the oscillations of a compass needle placed in the earth's field.

The main magnetic field of the earth. The north magnetic pole of the earth is that point on the earth where the north-seeking pole of a compass needle points straight down (vertically). This is also called the *dip pole*. At the south magnetic pole of the earth, the south-seeking pole of the compass needle points straight down. The positions of the north and south magnetic poles are constantly wandering. At the present time the north magnetic pole is near latitude 73 degrees north, longitude 100 degrees west in the Canadian Archipelago, while the south magnetic pole is near latitude 68 degrees south, longitude 146 degrees east on the edge of Antarctica.

We may visualize the magnetic field of the earth as corresponding to that which would be produced by a bar magnet placed inside the core of the earth. To produce the observed magnetic field, we would have to tilt the magnet about 11 degrees from the earth's rotational axis, at the present time, and place the center of the magnet some distance from the center of the earth. The field thus produced is called the *dipole* field (not to be confused with the dip pole). The magnetic north and south poles corresponding to this equivalent dipole (called the geomagnetic poles) do not come out at exactly the positions of the actual magnetic poles.

A compass needle tends to align itself parallel to the magnetic lines of force. Since the magnetic poles do not coincide with the geographical poles, the magnetic lines of force are not arranged, in general, parallel to the geographical meridians. Thus the compass needle does not, in general, orient itself parallel to the meridians, except in certain locations. The angle between the compass needle (or lines of force) and the geographic meridian is called the magnetic *declination*. (The declination is sometimes also called the variation.) East declination means that the compass needle points to the east of geographic (true) north, and west declination means that it points to the west of geographic north. If you use a magnetic compass to determine the direction of north, you must correct for the declination. For example, the magnetic declination at New York is about 10 degrees west. Thus in New York, true north lies about 10 degrees to the east of (to the right of, or clockwise from) magnetic north. In the vicinity of the magnetic poles, a magnetic compass is almost useless because of the large variations in declination around the poles.

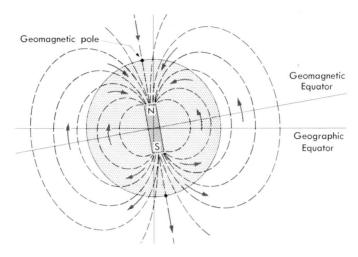

Fig. 34. Magnetic lines of force for a simple dipole model of the earth. Arrows show the orientation of the dip needle.

A compass needle which rotates in a vertical plane to indicate the tilt of the magnetic lines of force is called a *dip needle*. The angle between the dip needle and the horizontal is called the *inclination*. At the magnetic poles the inclination is 90 degrees, i.e., the dip needle points vertically. Where the magnetic lines of force are horizontal, the inclination is zero. The circle of zero inclination around the earth is called the magnetic equator. Figure 34 illustrates the magnetic lines of force (highly simplified) for a simple dipole field, and its effect on dip.

The magnetic field has two components, one horizontal, the other vertical. At the dip poles the horizontal field intensity is zero, and at the magnetic equator the vertical field intensity is zero. The intensity of the vertical field at the poles is about 0.63 gauss, which is about twice as large as the intensity of the horizontal field at the magnetic equator. Both components are measured with magnetometers at geomagnetic observatories, and on geophysical expeditions.

Variations in the earth's magnetic field. Because of the presence of magnetic materials in the earth's crust (e.g., magnetite, an iron oxide) the magnetic field exhibits some irregularities, or anomalies. These anomalies are useful in geophysical prospecting for mineral deposits in the earth. Airborne magnetometers are used to carry out geophysical surveys of otherwise inaccessible areas.

The magnetic field of the earth is constantly changing. The variations of the field are of three kinds: (a) slow, long-period changes, called secular variations, (b) regular, short-period oscillations, and (c) irregular, transient fluctuations.

The secular variations, including the wandering of the poles, are probably caused by motions within the earth's core, the same region that is probably responsible for the main magnetic field itself. There is now considerable evidence that the earth's magnetic field has changed very markedly with time. Such evidence is provided by the existence of "fossil" magnetism (paleomagnetism), the residual magnetism found "frozen" in rocks. The direction and intensity of this magnetization are very different from what we would expect if the magnetic field of the earth had always been oriented as it is today.

Paleomagnetism is still a confusing phenomenon, but it appears to indicate that the crust of the earth has moved relative to the core in the geologic past. It also appears to support the theory of continental drift (see p. 48) first proposed in 1912 by the meteorologist A. Wegener. Noting how well the Atlantic coasts of Africa and South America seem to fit together, Wegener suggested that they once were joined. This theory, that the continents may have migrated relative to each other, was rejected for many years by most geophysicists. However, with increasing evidence that the earth is less rigid than it was once thought to be, the theory of continental drift has begun to gain adherents. Significant support for the theory has been provided by paleomagnetism. The orientation of the ancient magnetic fields found "frozen" in the rocks of different continents is such that it apparently can only be accounted for by supposing that the continents have indeed moved relative to each other. The theory is, however, still highly controversial. For example, it has yet to be demonstrated that the theory accounts for the structure of the ocean floor.

Regular, short-period oscillations that are of small amplitude compared with the main field are observed on the magnetograms recorded at geomagnetic observatories. It has been established that these oscillations are caused mainly by electrical currents in the region of the upper atmosphere known as the ionosphere.* This is the layer of charged (ionized) particles which reflects radio waves back to the earth. The ionization is produced by ultraviolet radiation. The motion of these charged particles represents an electric conductor moving through a magnetic field, thereby inducing an electric current and a corresponding magnetic field. Thus the ionosphere makes a small contribution to the earth's magnetism. The periodic oscillations in this small contribution are caused by tides in the atmosphere (due to the sun and, to a lesser extent, to the moon) which produce periodic fluctuations in the ionospheric and in the magnetic field. In addition, there are regular variations caused by the fact that the amount of ultraviolet light received by the ionosphere from the sun varies with season and sunspot activity.†

---

* See Chapter IV, Section 6.
† See Chapter IV, Section 5.

Apart from the periodic variations, magnetic records also show irregular disturbances which begin suddenly and gradually die out in a matter of days. These are known as magnetic disturbances and magnetic storms. At times the disturbances are mild, and only localized effects are observed. At other times they are severe, and the whole earth is affected. The intensity of these storms increases from low to high latitudes. In general, the horizontal component of the magnetic field strength is most strongly disturbed, with variations amounting to about 0.05 percent of the total field strength. It is almost certain that these magnetic disturbances and storms are caused by events on the sun. This conjecture is supported by the fact that there is a close connection between the mean annual sunspot number (an indicator of solar activity) and the annual mean of the magnetic activity. Apparently, however, it is not the sunspots themselves which cause the magnetic activity, but some other manifestation of solar activity related to the sunspots. Which of the solar phenomena is directly responsible for the magnetic disturbances is not yet known with certainty.

The pattern of geomagnetic activity tends to recur in a 27-day cycle, indicating that there are definite regions on the sun, tentatively called M-regions, which contain the sources of geomagnetic activity. Because of the solar rotation, with its period of 27 days (making allowance for the earth's motion around the sun), these regions will face the earth every 27 days. Since the regions are not permanent but develop and disappear, magnetic storms do not exhibit a strict 27-day periodicity, but only a recurrence tendency, the actual recurrence depending on whether or not a specific M-region lasts through the whole solar rotation.

While the nature of the M-regions is not known, it is fairly well established that they emit charged particles which travel from the sun to the earth in one to four days, with a mean speed of 450 to 1750 kilometers (270 to 1100 miles) per second. It seems that solar flares, which are sudden violent solar eruptions, emit material which travels to the earth with such a speed and cause the great, suddenly commencing storms. As the electrically charged particles approach the earth and enter its magnetic field, they are deflected by it and spiral down toward the atmosphere around the magnetic poles, causing the greatest magnetic disturbances in polar regions. At the same time, they give rise in the upper atmosphere to *auroras,* which are thus closely connected with the magnetic storms.* †

---

* See Chapter IV, Section 6.
† See references 2, 15, and 16.

References

1. J. T. WILSON, *"The Crust,"* in *The Earth and Its Atmosphere,* D. R. Bates, ed. New York: Basic Books, 1957.
2. J. A. JACOBS, R. D. RUSSELL, and J. T. WILSON, *Physics and Geology.* New York: McGraw-Hill, 1959.
3. K. KRAUSKOPF, *Fundamentals of Physical Science.* New York: McGraw-Hill, 1959.
4. A. N. STRAHLER, *Physical Geography.* New York: Wiley, 1960.
5. S. N. NAMOWITZ and D. B. STONE, *Earth Science: The World We Live In.* Princeton, N.J.: D. Van Nostrand, 1953.
6. M. KAY, "The Origins of Continents," *Scientific American,* **193,** 62–66 (September 1955).
7. B. C. HEEZEN, "The Rift in the Ocean Floor," *Scientific American,* **203,** 98–110 (October 1960).
8. V. V. BELOUSEV, "Experimental Geology," *Scientific American,* **204,** 96–106 (February 1961).
9. W. L. STOKES, *Essentials of Earth History: An Introduction to Historical Geology.* Englewood Cliffs, N.J.: Prentice-Hall, 1960.
10. L. D. LEET and S. JUDSON, *Physical Geology,* 2nd. ed. Englewood Cliffs, N.J.: Prentice-Hall, 1958.
11. J. L. KULP, "Geologic Time Scale," *Science,* **133,** 1105–1114 (April 1961).
12. K. E. BULLEN, "The Deep Interior," in *The Earth and Its Atmosphere,* D. R. Bates, ed. New York: Basic Books, 1957.
13. B. F. HOWELL, JR., *Introduction to Geophysics.* New York: McGraw-Hill, 1959.
14. J. B. MACELWANE, *When the Earth Quakes.* Milwaukee: Bruce, 1947.
15. E. H. VESTINE, "Geomagnetic Field," in *The Earth and Its Atmosphere,* D. R. Bates, ed. New York: Basic Books, 1957.
16. S. CHAPMAN, *The Earth's Magnetism.* New York: Wiley, 1951.
17. J. H. HODGSON, *Earthquakes and Earth Structure.* New York: Prentice-Hall, 1964.
18. J. TUZO WILSON, "Continental Drift," *Scientific American,* **208,** 86–100 (April 1963).
19. A. N. STRAHLER, *The Earth Sciences.* New York: Harper & Row, 1963.
20. EARTH SCIENCE CURRICULUM PROJECT, *Investigating the Earth.* Boulder, Colo.: ESCP, American Geological Institute, 1965. (Text and Laboratory Manual.)

III. The Hydrosphere

1. The hydrologic cycle. Water is one of the few substances which is normally found on earth in all three phases: solid (ice), liquid, and gas (vapor).

In its solid form, water is present in the atmosphere as snow, and as ice crystals in clouds. It appears on the earth in the form of snow fields, frozen water in soil, and as continental and alpine glaciers. Ice is also present in what is commonly referred to as the hydrosphere—in the form of icebergs and sea ice in the oceans, and as frozen water in lakes, streams, and rivers. Ice interacts with the earth, the sea, and the air. In the form of glaciers, it alters the shape of the land. It emerges from the atmosphere by sublimation, from the hydrosphere by freezing, and returns by melting and evaporation. It cools the air and the sea, and is in turn heated by them.

More than 70 percent of the earth's surface is covered by the waters of the oceans, the principal reservoirs of liquid water on earth. Water is also found in the air in the form of clouds and falling rain, and on and under the earth in lakes, streams, rivers, and underground waters, as well as in the biosphere. It, too, interacts with the air and the land, altering land forms by erosion and transport of soil, leaving the atmosphere by condensation and returning to it by evaporation, and transferring energy from the sea to the air.

Water vapor is the invisible gaseous form of water, and the form in which it exists principally in the atmosphere. Water vapor enters the atmosphere by evaporation of seawater and freshwater, and by transpiration from plants. It leaves the air when it condenses in the form of dew, frost, or cloud particles, and precipitates out of the air as snow or rain.

Pure water is chemically unaltered by these changes of phase. It remains H_2O, whether it be vapor, liquid, or ice. What is altered by the phase changes is the cohesion of the molecules and the density of the water. A water molecule consists of two atoms of hydrogen bound to an atom of oxygen by an interatomic electrostatic attraction known as the *covalent bond*. These H—O—H molecules are in turn joined to other water molecules by an intermolecular electrostatic force known as the *hydrogen bond*. The nature and strength of the hydrogen bond determines the physical character of the water. In ice crystals the water molecules are bound to each other in rigid patterns, but with large spatial separation between the molecules. In liquid water the geometrical arrangement of the molecules is less rigid, and the molecules are closer together. As a result ice is less dense than liquid water. In water *vapor* the arrangement of the molecules is even more random and "loose." Furthermore, an increase of temperature of the vapor, which represents increased motion and kinetic energy of the molecules, separates the molecules still more and lowers the density of the vapor.

The change of phase from liquid water to water vapor, known as evaporation, requires the expenditure of a certain amount of energy or heat, called the *heat of vaporization*. Thus when seawater evaporates, the heat required for the process is taken from the air or the sea, which are thereby cooled. This heat remains dormant in the water vapor in a form called *latent heat of condensation*. When the water vapor condenses, the latent heat of condensation is given up by the water since its molecules are bound more closely in liquid form. The latent heat then warms the air. Thus evaporation takes heat away from the environment, and condensation returns it.

The phase change from water vapor to ice is called *sublimation*. The opposite change, from ice to vapor, is also called sublimation, although to avoid confusion we sometimes call it (erroneously) evaporation. Somewhat more energy is needed to convert ice to vapor than to convert liquid water to vapor. When the vapor again sublimes to ice, as in the formation of frost, the latent heat of sublimation is given up, and the air is warmed.

The melting of ice to form liquid water consumes an amount of energy known as the *heat of fusion*. While locked dormant in the liquid phase, this heat is called *latent heat of fusion*. It is released when the liquid water freezes. Thus, melting of ice, like evaporation of water, takes heat away, while freezing of water, like condensation of water vapor, returns it to the environment.

The transformation of water through all its phases on the earth is known as the *hydrologic cycle* (illustrated schematically in Fig. 35).

Water is transformed into vapor principally by evaporation from the ocean surface, but also by evaporation of freshwater on land and by transpiration from plants. The water vapor in the atmosphere is transported by winds over great distances, and may be carried from sea to land or from land to sea. As the rising water vapor is cooled, it condenses to form clouds of liquid water and ice crystals. The cloud drops and crystals grow until they fall to the earth as precipitation (rain and snow). Some of the precipitation feeds the snow packs in the mountains and polar regions, and contributes to the growth of glaciers. Most of it falls back into the ocean. Of the precipitation that falls on land, some enters the biosphere through plants and animals. Much of it flows over the land as runoff to join the streams that, together with snowmelt, feed the rivers which ultimately pour their waters into the sea. All freshwater on earth is provided by precipitation. Some of it infiltrates into the soil, where it contributes to the groundwater supply and to the underground streamflow. The rest returns to the atmosphere by evaporation.

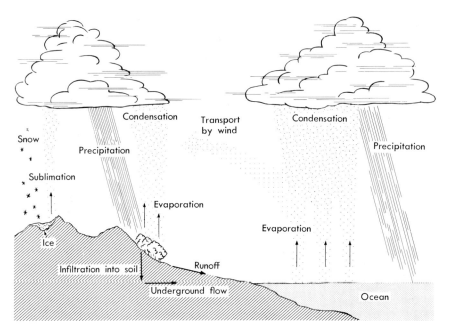

Fig. 35. The hydrologic cycle.

The hydrologic cycle is closed. The same water recirculates constantly through all the phases. For the earth as a whole, the total amount of water precipitated on the earth each year is, on the average, equal to the total amount of water that evaporates each year from the earth's surface. The amount of water permanently stored in the atmosphere is only a small fraction of this annual evaporation and precipitation, and does not change much from year to year. The average annual rainfall of the earth, which is also the average annual evaporation, is about 75 centimeters (30 inches). This corresponds to more than a hundred million billion (10^{17}) gallons of water exchanged each year between the earth and the air.

Although the annual world precipitation nearly equals the annual world evaporation, this is not true of individual localities on the earth. Thus over some regions of the oceans, rainfall exceeds evaporation, whereas in other regions evaporation exceeds rainfall. This pattern of excess and deficit of water is compensated for by the circulation of the oceans, both in surface and deep-water currents, so that the sea level tends to remain nearly unchanged with time.

One important long-term factor in the world hydrologic cycle is the water locked up in the glaciers. The formation of glaciers lowers the sea level, and the melting of the ice raises it. At the present time, 10 percent of the surface of the continents, or about three percent of the earth's total surface, is covered by glaciers. Should all this ice melt and be spread over the oceans, it would form a layer of water about 57 meters (186 feet) thick. However, the sea level would rise by only about 38 meters because of isostatic adjustment in the earth's interior.

During the ice ages the sea level was lower (perhaps by more than 30 meters) than it is today, as a result of the storage of water in the great continental ice sheets. Some scientists believe that the lowering of sea level finally put an end to the ice age by cutting off the exchange of water that takes place across a shallow sill between the Arctic Ocean and the Atlantic. This event, they believe, led to a warming of the Atlantic, and pushed the zone of heavy snowfall farther north, thus allowing the melting and retreat of the continental glaciers in middle latitudes.

The possible effect of the ocean circulation on long-term climatic changes is but one example of the importance of the oceans in controlling and regulating world climate through their interaction with the atmosphere. We shall next study some of the geographical and physical characteristics of these vast bodies of seawater, which play such an important role in all aspects of life on earth.*

2. The ocean basins. *The breadth and depth of the seas.* The oceans cover 71 percent of the earth's surface. In the Northern Hemisphere, 61 percent of the earth's surface is ocean, and in the Southern Hemisphere the oceans occupy 81 percent of the surface area. Figure 36, an uninterrupted, equal-area map of the oceans, illustrates the great extent of the oceans, especially in the Southern Hemisphere.

The term "ocean" is used here to refer to not only the three principal oceans—the Atlantic, Pacific, and Indian Oceans—but also the adjacent seas. (The student should consult a map, a globe, or atlas to familiarize himself with the geography of the earth in general, and with the oceans in particular.)

The large mediterranean (between the land) seas are the Afro-European Mediterranean (the ancient Mare Nostrum), the Black Sea, the Arctic Ocean, the Asian "mediterranean," which includes the Sea of Okhotsk, the Japan Sea, and the China Seas, and the so-called American "mediterranean," which consists of three basins: the Gulf of Mexico, and the eastern and western Caribbean Seas.

The small mediterranean seas include the Baltic Sea, Hudson's Bay, and many others. Like the large mediterranean seas, these communicate to the oceans through relatively narrow passages (e.g., the Straits of Gibraltar) or over relatively shallow barriers called *sills*. In addition, there are the so-called marginal seas, such as the Bering Sea and the North Sea, which lie along the continental borders of the oceans. These are generally deeper than the shallow small mediterranean seas.

---

* Reference 1.

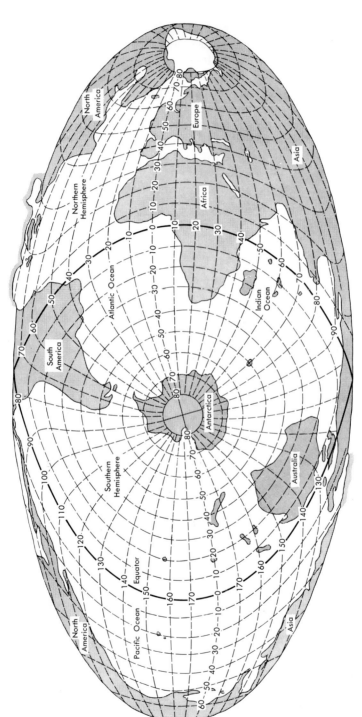

Fig. 36. Uninterrupted equal-area map of the oceans.

TABLE 6
AREAS, WATER VOLUMES, AND AVERAGE DEPTHS
OF THE OCEANS AND SEAS

Region	Surface area, million square kilometers (miles)	Volume, million cubic kilometers (miles)	Average depth, meters (feet)
Atlantic Ocean	82 (32)	323 (78)	3930 (12,900)
Pacific Ocean	165 (64)	708 (171)	4280 (14,000)
Indian Ocean	73 (28)	291 (70)	3960 (13,000)
Large mediterranean seas	30 (12)	41 (10)	1380 (4,400)
Small mediterranean seas	2 (1)	0.4 (0.1)	170 (560)
Marginal seas	8 (3)	7 (2)	870 (2,850)
Total	360 (140)	1370 (331.1)	3800 (11,600)

TABLE 7
PERCENT OF THE EARTH'S SURFACE COVERED BY WATER
IN DIFFERENT LATITUDE BANDS

Northern Hemisphere		Southern Hemisphere	
Latitude, °N	Percent water	Latitude, °S	Percent water
90–70	75	70–90	20
70–50	37	50–70	96
50–30	53	30–50	92
30–10	68	10–30	77
10–0	77	0–10	76

The surface areas, water volumes, and average depths of the principal water bodies of the earth are shown in Table 6. Note that more than half the water (volume) of the earth is in the Pacific Ocean.

The relative proportion of land and water varies with latitude, as shown in Table 7. Note the high percentage of water in middle latitudes of the Southern Hemisphere compared with the corresponding latitudes in the Northern Hemisphere.

Depth measurements. The depths of oceans are measured today with echo-sounding devices from oceanographic research vessels. Prior to the development of the echo sounder, the ocean depths were measured by rope and, later, wire soundings in which a heavy weight is paid out on a thin steel wire over a measuring wheel until it reaches bottom. (On an expe-

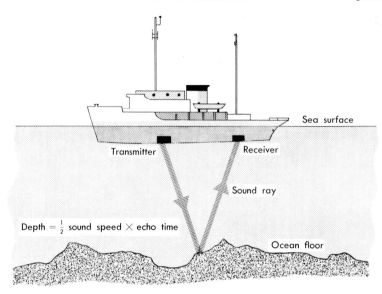

Fig. 37. Determination of ocean depth and bottom topography of the sea by the echo-sounding method.

dition to the Moluccan Seas near Indonesia the Dutch research ship *Willebrord Snellius* measured a depth of 10.2 kilometers (33,500 feet) with a wire sounder in the Mindanao Trench off the Philippines.) Modern oceanographic research vessels still carry wire and winches for soundings, for dredging the sea bottom for samples of the earth's floor, and for bringing up core samples of oceanic sediments. Some of these vessels carry 12 kilometers (40,000 feet) of wire, enough to reach the bottom of the deepest known trenches, which extend more than 10.7 kilometers (35,000 feet) below sea level. The pull on these wires when they are loaded and at the bottom of the sea may exceed 12 tons.

The echo-sounding method, illustrated in Fig. 37, is based on the speed of sound in water. A short, sharp sound signal is emitted by the sounding vessel while it is underway. The sound waves, which travel in all directions, are reflected from the sea bottom back to the ship. The time interval between the emission of the sound and the return of the echo from the sea floor is measured and recorded on a chart. The depth of the sea is equal to the speed of sound in water multiplied by the time interval and divided by two. (The sound wave travels a total distance equal to twice the depth.) For example, the speed of sound in seawater is approximately 1500 meters per second. If 10 seconds elapse between transmission and return of the sound, the sound wave must have traveled 15 kilometers. Since the sound must travel down and back, the depth of the sea is only half this distance, or 7500 meters.

Because water is so much less compressible than air, the speed of sound in water is about five times as fast as the speed of sound in air, and is faster in seawater than in pure water, an effect of the salinity. Sound does not travel with constant speed through the sea. It travels faster in warm water than in cold water. The speed of sound tends to increase with increasing depth in the ocean due to the increase in pressure. These effects of salinity, temperature, and pressure on the speed of sound in seawater are taken into account in the echo-sounding method.

Echo sounding may also be used in reverse, to determine the height of the bottom of the Arctic pack ice (sea ice) above a submarine, as shown in Fig. 38. Sound ranging (called sonar) is also used to measure distances to floating objects, such as enemy vessels.

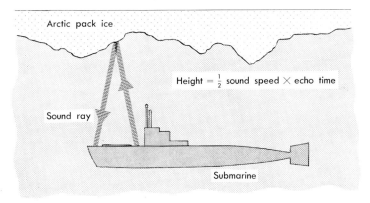

Fig. 38. Determination of the height of the bottom of the Arctic ice pack above a submarine.

The depth of the ocean may also be determined by measuring the pressure. The pressure at any depth in the ocean is equal to the weight of a vertical column of ocean water divided by the cross-sectional area of the column, i.e., the weight per unit area. The weight of a column of water per unit area depends mainly on the height (depth) of the water. (It also depends on the density of the water, but the variations in water density are relatively small.) Thus the pressure is almost exactly proportional to the depth of the water. Atmospheric pressure at sea level (the weight per unit area of the atmosphere at sea level)* is approximately equivalent to the pressure of a column of water 10 meters deep. Thus the pressure in the ocean increases by one *atmosphere* for every 10 meters of depth. At the bottom of the deepest trenches in the ocean, the pressure is more than *one thousand* times the pressure at sea level.

---

* See Chapter IV.

The pressure at great depths can be measured with two thermometers, one of which is protected from the ocean pressure, while the other is not. The difference between the temperature readings of the protected and unprotected thermometers is a measure of the pressure, and hence of the ocean depth to which the thermometers have been lowered.

The principal features of the ocean bottom. From wire soundings and the much more numerous echo soundings we are now beginning to develop a reasonably complete picture of the configuration of the ocean floor.

Except for some ridges and isolated peaks, the sea floor is covered by sediments (see p. 50). Extensive seismic surveys of the sediments appear to indicate that their average thickness throughout the deep oceans ranges from 0.5 to 1 kilometer. Samples of the uppermost layer of these thick sediments have been obtained by means of core samplers (or core barrels). These are cylinders forced into the sediments by dropping heavy weights on them. It is extremely difficult by this method to obtain core samples more than 30 meters thick. But because the sea bottom is relatively unaltered by geological processes, even cores of this thickness may encompass more than 25 million years of geological, biological, and climatological history.

The features of the submarine topography resulting from deformations of the earth's crust are either elevations or depressions. The former include ridges, rises, seamounts, and sills, and the latter include troughs, trenches, basins, and deeps. Features resulting from erosion, deposition, and biological activity include the continental shelf, the continental slope, banks, shoals, reefs, canyons, and valleys. (See Chapter II, Section 2 for further details.)

The Atlantic Ocean is divided into two troughs by a north-south ridge, the Mid-Atlantic Ridge. On either side of the ridge, depths exceeding 5 kilometers exist, although the depth of the ridge itself is less than 3 kilometers, and in some places (islands) it comes to the surface. The ridge extends into both hemispheres.

A similar north-south ridge bisects the Indian Ocean south of India between Africa and Australia. It extends all the way to the Antarctic continent, and is wider and deeper than the Atlantic ridge.

In the Pacific Ocean the ridges are broken up and are less conspicuous. There is one ridge from Japan to Antarctica, and another from Central America southwestward to Antarctica. In the central and western Pacific there exists a complicated system of ridges and troughs with no clear-cut longitudinal orientation.

Both the oceans and the mediterranean seas consist of a collection of basins separated by sills. Thus the so-called "Arctic mediterranean" is composed of a deep broad central area, the North Polar Basin, and a number of smaller basins. According to recent Russian observations, it seems that the North Polar Basin, once thought of as a single elliptical basin with an average depth of about 4 kilometers in its central area, is actually bisected by a ridge, the Lomonosov Ridge.

In the Antarctic, three or more distinct basins surround the Antarctic continent. The bottom topography of these basins is still being explored.

Topographic maps of the ocean bottom are called bathymetric charts. Contour lines drawn on these maps to show the depth of the ocean floor are called *isobaths* (lines of equal depth). Isobaths may be labeled in meters (1 meter = 3.3 feet), or fathoms (1 fathom = 6 feet). The isobaths are crowded together where the bottom slopes steeply, as on the edge of a trench, trough or ridge, and are far apart where the ocean floor is flat. Bathymetric maps of the oceans are published by governmental agencies, such as the United States Navy Hydrographic Office. Standard bathymetric charts of the oceans are published by the International Hydrographic Bureau in Monaco.*

3. The oceans. *Physical properties of pure water.* To understand the physical processes that take place in seawater, we shall first review the physics of pure (fresh) water, and then see how the presence of salt affects the behavior of the water.

The freezing-melting point of water is the temperature at which water and ice are in equilibrium. (To produce this temperature, drop ice cubes into a pail of water until the ice cubes stop melting.) By definition, this temperature is zero on the celsius (formerly called centigrade) scale, i.e., 0°C. It is also, by definition, 32 degrees on the fahrenheit scale, i.e., 32°F.

The boiling point of water is by definition 100°C and 212°F, at standard pressure. There are 180 degrees (212° to 32°) on the fahrenheit scale corresponding to 100 degrees on the celsius scale, or a ratio of 9°F for every 5°C. To convert fahrenheit temperatures to celsius, we subtract 32 from the fahrenheit temperature and multiply this difference by $\frac{5}{9}$. (Example: Show that 41°F is equal to 5°C.) To convert celsius temperatures to fahrenheit, we multiply the celsius temperature by $\frac{9}{5}$ and add 32 to the result. (Example: Show that 10°C is equal to 50°F.) (Another scale of temperature, the absolute, or kelvin, scale, is constructed by adding 273 to the celsius temperature.)

The density of pure water is almost exactly one gram per cubic centimeter (1 gm·cm^{-3}). Although water is almost incompressible (compared, for example, with air), its density does change slightly with temperature. For example, if we cool water from a temperature of 20°C to a temperature of 4°C, the water shrinks, and its density rises from 0.99821 gm·cm^{-3} to 0.99997 gm·cm^{-3}. But if we continue cooling the water below 4°C, it expands again and its density decreases. For example, at 0°C the density

---

* See references 1 through 8.

of water falls to 0.99984. Thus water has a *temperature of maximum density* equal to 4°C.

If water is cooled below 0°C, some of it will freeze. While freezing the water expands noticeably. (Water freezing in a glass bottle will expand enough to shatter the container.) The density of ice is only about 92 percent of that of liquid water. Consequently, an ice cube will float in a glass of water, although about $\frac{9}{10}$ of the cube is submerged. (If we pour alcohol into the glass, the cube will sink because the density of alcohol is less than the density of ice.) The buoyancy force which supports a submerged object against the pull of gravity is, according to Archimedes' (287–212 B.C.) principle, equal to the weight of the fluid displaced by the object. For a floating object the fraction of the volume submerged is equal to the ratio of the density of the object to the density of the fluid in which it floats.

When a freshwater body, such as a lake, cools in winter, the surface-water density increases until a temperature of 4°C is reached. The heavy, cold water sinks as it is cooled, and is replaced by lighter, warmer water from below. This process continues until the entire lake has cooled to a uniform temperature of 4°C (39°F). Further cooling of the surface water causes it to expand. This light, cold water now remains on the surface of the lake, and freezes there after the temperature has fallen below 0°C. Thus a coating of ice forms on the surface of the lake.

Properties of seawater. Seawater differs from pure water primarily because of the presence of dissolved salt in the water. (Undissolved suspensoids and biological organisms also contribute to the difference.) The salt in the sea, which is a product of the weathering of rocks, is being added to constantly by the rivers, which carry dissolved minerals into the sea. Evaporation from the sea surface removes pure distilled water, leaving the salt behind. It is not known with any certainty how rapidly the salt content of the oceans has actually increased during geological history as a result of this process, for salt also leaves the sea to form salt deposits.

The *salinity,* or amount of salt dissolved in the ocean water, is measured chemically. Oceanographers collect water samples from the sea surface and from great depths in bottles or in metal containers called sea samplers. In shipboard laboratories (or back on land) they perform quantitative chemical analyses to measure the chlorine content (chlorinity) of the water. Chlorine ions constitute about 55 percent of all the dissolved material in seawater, and the total salinity can be determined from the chlorine content. The salinity can also be determined by measuring the electrical conductivity of the seawater sample, for the conductivity depends upon the amount of salt dissolved in the water. The electrical method is widely used on oceanographic research vessels today.

Salinity is expressed in terms of the concentration of salt, i.e., the mass of salt in a unit mass of seawater, and is usually expressed in parts per thousand, called "per mille" and written ‰. (Compare with percent, meaning parts per hundred and written %.) The average salinity of seawater is about 35 per mille, i.e., there are about 35 grams of salt dissolved in every kilogram of seawater. (To evaporate this quantity of water would require an expenditure of energy, for the heat of vaporization alone, equivalent to burning 70 grams of coal. The economics of exchanging 70 grams of coal for 35 grams of salt are questionable. However, seawater can be evaporated and recondensed to yield fresh water, and salt can be recovered economically by the use of solar energy for evaporation.)

The density of seawater increases with increasing salinity, and is, on the average, about two to three percent greater than that of freshwater. Two effects of the increased density of seawater are greater buoyancy, which is familiar to every swimmer, and a tendency for salt water to sink below and form a wedge underneath the lighter freshwater in estuaries, where the rivers meet the sea. (This happens, for example, in the Hudson River near New York City.) The increased buoyancy of seawater causes icebergs to float higher in the ocean than they would in freshwater. Thus, unlike the ice cube in the water glass, something *less* than nine-tenths of an iceberg is below the water line. Indeed, because of the air trapped in the ice, as little as six-tenths to eight-tenths of the iceberg may be below the water line.

The freezing point of water decreases as the salinity increases. Seawater with a salinity of 25 per mille (e.g., estuarine water) has a freezing point of $-1.3°C$, whereas seawater of average salinity (35 per mille) has a freezing temperature of about $-2°C$. This temperature, $-2°C$ or about $28°F$, is the lowest water temperature found anywhere in the sea. Freshwater bodies obviously freeze more readily than seawater.

The temperature of maximum density of seawater also decreases as the salinity increases. Thus at a salinity of 25 per mille the temperature of maximum density is equal to the freezing point, $-1.3°C$. At higher salinities such as are found in the ocean, the temperature of maximum density is lower than the freezing point. This means that the density of seawater, unlike that of freshwater, increases as the temperature falls, all the way down to the formation of sea ice. The light (low-density) water in the sea is the water with high temperature and low salinity. The heavy (high-density) water is the cold, saline water.

The distribution of temperature and salinity. The temperature of the surface water of the ocean is easily measured by ships using ordinary thermometers. Temperature may be measured at the condenser (engine) intake of the ship, although this method often results in erroneous readings. It is more satisfactory to measure the sea temperature in bucket samples of seawater. Oceanographic-research vessels may use special thermographs. A recently developed method makes it possible to determine the surface temperature of the sea from an airplane, by measuring the infrared radiation emitted by the sea surface.

The temperature of the sea *below* the surface may be measured with a *bathythermograph* (Gr. *bathos,* depth, plus temperature), an instrument which records the change of temperature with pressure on a smoked plate. The record of temperature versus pressure obtained with a bathythermograph is equivalent to a record of temperature versus depth. The bathythermograph is ordinarily used only for relatively shallow soundings (down to about 1000 feet). For greater depths *reversing thermometers* are customarily employed. The latter, like the bathythermograph, are lowered on a wire to predetermined depths. They are then triggered by dropping a weight, called a "messenger," along the wire. This causes the thermometers to turn over (reverse), breaking the thread of mercury in the special bore of the glass-thermometer tube, and fixing the length of the mercury column. The thermometers are then hauled up to be read.

The warmest water in the open ocean is found near the Equator, where the temperature averages 28°C (82°F) and is occasionally as high as 30°C (86°F). The coldest water, which is found in the Arctic, has a temperature of −2°C (28°F).

The temperature of surface water of the deep oceans changes little with the seasons, and hardly at all from day to night. This absence of pronounced variations is due to the fact that heat added to or taken from the ocean is distributed by the motion of the sea over great masses of water, giving the ocean a large heat capacity. The ocean acts as a heat regulator for the atmosphere, and reacts sluggishly to changes of air temperature. In shallow water, e.g. on the continental shelf, the sea-surface temperature changes more markedly with season and weather conditions than in deep water.

The temperature of the sea typically decreases with increasing depth below the sea surface. (However, one frequently encounters shallow surface layers in which the temperature increases with depth.) There is usually a layer at moderate depths, in which this change of temperature with depth is at a maximum. This layer, called the *thermocline,* is generally less than a thousand feet thick. Below the thermocline, the temperature of the sea falls only very slowly with increasing depth, and below a depth of 1500 meters the temperature is almost constant. In this nearly isothermal deep water, the temperature is between one and three degrees celsius.

In addition to the variation in sea temperature from Equator to pole, which is caused by the variation of solar radiation with latitude, the sea temperature also varies from east to west, especially near the continents. Thus we find warm currents of water, such as the Gulf Stream in the Atlantic and the Kuroshio Current in the Pacific, moving poleward, and cold currents, such as the Canary Current in the Atlantic and the California Current in the Pacific, moving toward the Equator. The variations in temperature from east to west across these currents are generally much greater than the north-south temperature changes. (The cause of this temperature pattern is discussed below.)

The temperature distribution is represented by drawing lines of equal temperature, called isotherms. Along the edges of the cold and warm currents, where the temperature changes markedly with distance, the isotherms are crowded together. In the equatorial region, where the sea temperature is uniform, the isotherms are far apart.

The salinity of the surface water of the oceans depends upon the ratio of evaporation to precipitation (rain and snow). High evaporation and low precipitation lead to an increase in the salt concentration, whereas low evaporation and heavy precipitation reduce the salinity as a result of dilution by addition of freshwater. Hence, the salinity is low in the equatorial region, because this is the region of heaviest rainfall on earth. In the subtropical latitudes, about 25 degrees north and 25 degrees south, there is a minimum of rainfall, the skies are clear, and evaporation is high. Thus the subtropical ocean surface water has the highest salinity of any ocean water. As we move from the subtropics toward the poles, salinity decreases due to the increasing rainfall and decreasing evaporation. Salinity in the open ocean ranges from a minimum of 30 per mille to a maximum of 37 per mille.

In certain nearly enclosed basins where the evaporation greatly exceeds the precipitation, the salinity is remarkably high. Extreme examples of such basins are the Great Salt Lake in Utah and the Dead Sea near Jerusalem. In the latter, the salinity exceeds 200 per mille. Less extreme examples are the Red Sea, with salinities greater than 40 per mille, and the Mediterranean Sea. The Red Sea and the Mediterranean receive relatively little freshwater from rivers.

Since the density of seawater increases with increasing salinity, it follows that excessive evaporation increases the density of the surface water, which in turn causes the heavy saline water to sink. Thus evaporation is responsible for a vertical circulation (overturning) of the ocean water. The light fresh rainwater, on the other hand, tends to float at the surface of the sea. In general, the densest (heaviest) water is at the bottom of the ocean, with the lighter water resting on top of this bottom layer.

Gases and life in the sea. Life in the sea is dependent on the presence of dissolved oxygen in the ocean water. Animals in the sea consume oxygen which must be replaced by the oxygen produced by plants in the sea and by atmospheric oxygen. The total amount of oxygen in the sea is negligible compared with that in the atmosphere, so that the latter is a more than sufficient reservoir from which the ocean's oxygen demand can be replenished. However, there is an upper limit to the amount of oxygen that can be dissolved in the sea. When the surface water reaches this state of oxygen saturation, no further solution of atmospheric oxygen occurs.

The oxygen in seawater is used as an index of the biological history of the water, and also as an indicator of its circulation and mixing history. Water recently exposed to air tends to be rich in oxygen, whereas water that has not been in contact with air for a long time may be deficient in oxygen and unable to support life. In deep stagnant water, where overturning and water exchange are minimal, the decay of dead organic material may use up almost all the dissolved oxygen, thus leading to a high concentration of toxic hydrogen sulphide. No animal life is possible in this poisonous water until the water overturns and the oxygen is replaced.

The oxygen content of the sea is usually greatest near the surface. However, the vertical distribution of oxygen is rather complicated. In many places, layers of minimum oxygen are found at intermediate depths rather than in the deepest water. The existence of life and oxygen even in the deep regions of the sea indicates that there must be circulation and exchange of water between these great depths and the surface.

Since the overturning of the ocean water increases the total oxygen content, it also increases the abundance of life in the sea. Another way in which this oceanic convection enriches the sea, especially in shallow seas and fishing banks, is by bringing the nutritive products of decay to the surface where they feed the floating vegetation on the sea surface. This ultimately increases the abundance of fish in the sea.

There is much more carbon dioxide (CO_2) (about 60 times as much) in the sea than in the air. The capacity of the oceans to absorb carbon dioxide is very great, and hence the oceans act as regulators for the amount of carbon dioxide in the atmosphere. When the CO_2 content of the atmosphere increases, the rate at which the gas dissolves in seawater increases also, thus keeping the atmospheric concentration of CO_2 from rising excessively. The carbon dioxide in the sea is consumed by marine plants.

The sun and the sea. The distribution of temperature at the sea surface, from a maximum at the Equator to a minimum at high latitudes, is brought about by the global distribution of incoming solar radiation falling on the earth's surface. Due to the variation in the inclination of the sun's rays with latitude, *insolation* (an abbreviation for *in*coming *sol*ar radi*ation*) is greatest at the Equator, and smallest at the poles. The distribution of insolation varies with the seasons as the sun moves back and forth across the Equator.

Not all the insolation falling on the sea is used to warm the sea. A fraction of this energy is reflected by the sea surface back to space. This reflectivity, called the *albedo,* varies with the solar angle, being greatest when the sun is low in the sky. Another portion of the insolation is used for evaporation. This energy is realized in the form of heat only after the evaporated water has condensed in the air to form clouds and rain. Furthermore, the sea is at all times radiating energy to space in the form of invisible, long-wave, infrared radiation. The temperature of the sea is the net result of all these processes; however, it is also affected by the exchange of heat between the sea and the air, and by the transport of energy via ocean currents and convection.

It has already been noted that the heat capacity of the oceans is very great, largely because the turbulent motion of the oceans distributes the gains and losses of energy over vast masses of water.

That part of the solar and sky radiation which is not reflected penetrates into the sea and is absorbed there. The resulting loss of radiative energy is measured by the *extinction coefficient,* which varies with the wavelength (color) of the light and with the type of water. Coastal waters have a larger extinction coefficient (or higher turbidity) than waters farther out in the ocean. This is due to the presence of fine suspended matter, much of it of organic origin, of which there is more in coastal waters than in the deep ocean. Because of the extinction of light in the oceans, the visible light is reduced to 15 percent of its surface value at a depth of 50 meters, and to about 7 percent at a depth of 100 meters. Thus at great depths the oceans are quite dark.

The *color of sea water* varies in different parts of the oceans. It may be a deep blue or an intense green, or sometimes brown or brownish red. Blue waters are typical of the open oceans, particularly in middle and low latitudes, green is more common in coastal waters, brown or brownish-red hues are observed only in coastal waters.

The blue color is explained by the scattering of light by water molecules which are small compared with the wavelength of the scattered light. The intensity of the scattered light increases inversely as the fourth power of the wavelength. Hence the blue component of the sunlight, which has a short wavelength, is scattered more than, for example, red light, which has a longer wavelength. (The blue color of the sky is similarly caused by the scattering due to air molecules in the atmosphere. The blue skylight also contributes to the blue color of the sea.) Where larger particles, either organic or inorganic, are present in the oceans, the water does not have this blue color. It follows that the blue color indicates the absence of small-animal and plant life in the oceans, and blue has therefore been called the "desert color" of the sea. The green color is due to the admixture of the so-called "yellow substance," which is produced by the metabolic activities of animal and plant life. Great quantities of larger suspended particles may impart to the sea other colors (e.g., red and yellow).

Ice in the sea. There are two kinds of ice found in the sea: icebergs and sea ice.

Icebergs are produced when the edge of a glacier breaks off and falls into the sea. This process is called *calving*. The ice in an iceberg is completely fresh.

The two principal sources of icebergs in the world are the great continental glaciers of Greenland and Antarctica. As the glacier flows toward the coast (in Greenland this occurs mainly in the fjords), part of it breaks off and falls into the sea with tremendous violence. In spring the icebergs break away from the glacier and are carried by currents to warmer latitudes. The Greenland icebergs (more than 10,000 each year) drift southward into the Atlantic shipping lanes where they are watched carefully by the International Ice Patrol. Spring and summer are the seasons when the iceberg danger to shipping is at a maximum.

Greenland icebergs are generally smaller (less than 75 meters high and less than 600 meters long) than the monster Antarctic icebergs. The former rarely last more than two years, but the latter may survive for more than 10 years. As the iceberg melts away, it drops the enclosed moraine on the sea floor.

Most of the ice in the oceans, however, is produced by the freezing of seawater. This is called sea ice or pack ice. Although frozen seawater contains relatively little salt, there are many small cavities in this brittle, porous ice, and these contain either seawater or even saltier brine. However, over a period of time, this brine leaks out of the ice so that old sea ice may be melted down to produce relatively salt-free water.

Sea ice does not grow to very great thickness and rarely exceeds 3 meters, for the ice, once formed, acts as an insulator and protects the underlying water from excessive cooling. Where the sea is exposed to exchange with warmer oceans, as around the Antarctic continent, the ice disappears in summer. But in protected basins, such as the North Polar Sea, the ice is semipermanent, although it is thinner and develops cracks (leads) and holes in summer. Ice islands in the Arctic Sea have been known to last for several years. They drift continuously about the Arctic Sea during their lifetime and are therefore of considerable geophysical interest. Scientists of both the U.S. and U.S.S.R. have occupied ice islands in the Arctic in recent years for the purpose of carrying out geophysical explorations. As is well known, the breaking-up of the islands exposes these floating stations to considerable hazards.*

4. Currents, waves, and tides in the ocean. *Ocean currents.* The currents in the oceans have been observed since man first ventured to sail the seas. Careful measurements of ocean currents have been carried out by several methods. One method has been to observe the motion of floating objects, such as drifting wreckage from vessels (flotsam), jetsam, ice, and fisher-

---

* References 1 through 8.

men's implements. A more systematic and reliable method has made use of *drift bottles,* which are weighted with sand so that they float just below the water surface. More recently, *current meters* have been developed for measuring the flow of currents by means of propellers. However, since current meters must be held stationary relative to the earth, the meter is either supported on a tripod fixed on the ocean floor or suspended by a wire from an anchored ship. For the measurement of deep water currents, deep *drogues* may be employed. A drogue is any object with a large drag, such as a surplus parachute, which is suspended at any depth in the ocean by a fine wire attached to a surface float. The deep water current carries the drogue along, and its motion is determined by observing the movement of the float.

The *relative* ocean currents (that is, the motion of the water relative to a moving ship) can be observed by measuring the motion of objects thrown overboard, or by observing the movement of foam and whitecaps past the ship. However, the *absolute* current (the current relative to the earth or to a stationary object on the earth) can only be determined from a moving ship if the ship's absolute motion is known. With modern radio-navigational systems such as Loran, Shoran, etc. it is possible to determine a ship's position, and hence its absolute motion, very precisely. Thus absolute ocean-current measurements may now be made systematically by radio navigation in regions, such as the western Atlantic, where these systems are available.

Most of our information about the worldwide ocean current systems has been derived from the observed movements of ships. A ship's officer can determine the *average* ocean current experienced during a period of sailing by comparing his actual motion with that predicted from the compass heading of the vessel and its known speed in still water. The actual ship's motion may be determined by landfalls or from celestial (astronomical) navigational fixes. At the end of a day's sailing it is possible to calculate how the course of the ship was affected by the current. It was from thousands of current measurements like these, entered in ship's logbooks, that Mathew Fontaine Maury (1806–1873), an American naval officer, compiled his pioneer study of the ocean currents in 1855.

An ingenious, and now widely used, method for measuring absolute ocean currents, either from shore points or from a moving ship, is the *electromagnetic* method. This technique is based on the dynamo principle according to which an electrical current is generated when an electrical conductor cuts across the lines of force of a magnetic field. Seawater is an electrical conductor. Hence, the motion of an ocean current across the earth's permanent magnetic field produces a measurable electrical effect. Thus it is possible, for example, to determine the speed of the Florida current by measuring the difference in electrical potential (voltage) induced between Key West, Florida, and Havana, Cuba.

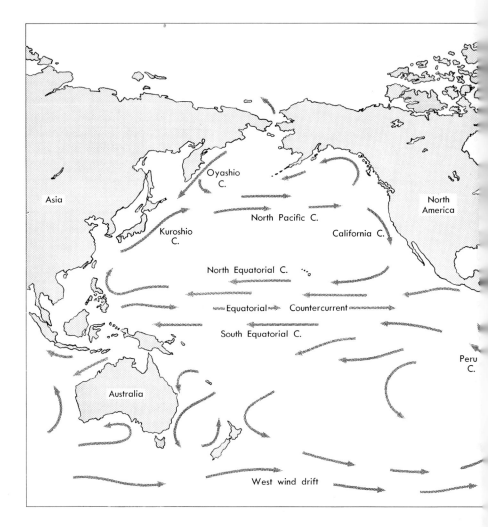

The electromagnetic method is used at sea by towing a long (300 foot) wire behind a ship. Attached to the wire are two floating electrodes. The electrical potential induced between the electrodes as the wire cuts across the magnetic field is recorded by means of a device known as a geomagnetic electrokinetograph, abbreviated GEK. The GEK records the effect of the sideways motion of the wire (and ship) as it is pushed by the component of the ocean current in a direction perpendicular to its heading. The ship is then, for a short time, turned 90°, at right angles to its original heading, to obtain a second component of the current. From the two measurements the absolute current can be computed. The GEK or towed-electrode method provides detailed information on the structure of ocean currents.

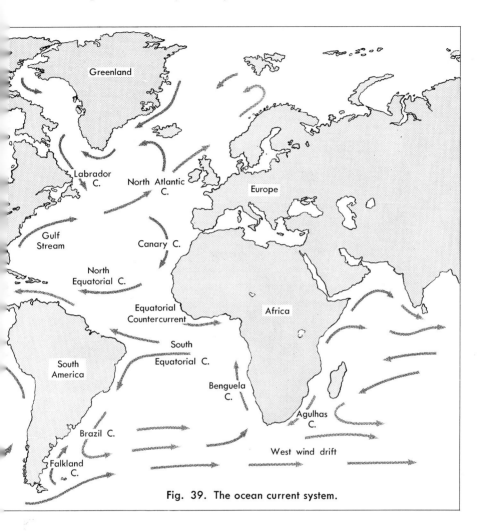

Fig. 39. The ocean current system.

One of the more recent current-measuring devices is the "Swallow float" (named after its inventor), which can be used to measure currents at or below the sea surface. The Swallow float is a cylindrical tube containing a battery-powered sonar transmitter, sometimes called a "pinger." The tube is weighted to float at any predetermined depth. The velocity of the current carrying the Swallow pinger is determined from the signal received from the transmitter.

The speed of ocean currents may be expressed in meters per second, in knots, in kilometers per day, or in nautical miles per day. A knot is equal to one nautical mile per hour. A nautical mile is equal to 1.15 statute (land) miles, and is also equal to one minute of latitude along a meridian. One degree of latitude is thus equivalent to 60 nautical miles, and to approximately 111 kilometers. It follows that one meter per second is equal to approximately 2 knots.

The worldwide system of surface ocean currents is illustrated schematically in Fig. 39. The outstanding features of the current system are the great whirls centered in the subtropical latitudes of the Atlantic and Pacific Oceans. The subtropical *gyres* rotate in a clockwise sense in the Northern Hemisphere and in a counterclockwise sense in the Southern Hemisphere. We shall see later that they are driven by corresponding wind systems.

On the western side of the North Atlantic subtropical gyre, we find the fast-moving (up to a maximum of five knots), poleward-directed, warm Gulf Stream. On the eastern side of this gyre, there is the equatorward-moving, cool Canary Current, which flows past the Canary Islands, off the coast of Africa. The southern limb of the gyre is the westward-flowing North Equatorial Current, while on the north side of the gyre flows the eastward-moving North Atlantic Current. Near the west central part of the gyre lies the Sargasso Sea, a region of clear, blue water containing little plant or animal life except for the characteristic seaweed (sargassum) that is carried into the region by the converging ocean currents.

A corresponding set of currents constitutes the North Pacific subtropical gyre. The warm Kuroshio Current (also known as the Japan Current) flows poleward on the western side of the Pacific, while the cold California Current flows equatorward on the eastern side of the ocean. As in the Atlantic, there is an eastward-moving current on the north side of the gyre (the Pacific Current) and a westward-flowing current (the North Equatorial Current) on the south side of the North Pacific gyre.

In the Southern Hemisphere the gyres are reversed. A current of relatively warm water, the Brazil Current, flows southward, away from the Equator, on the western side of the South Atlantic gyre. On the eastern side of the gyre, a current containing relatively cold water, the Benguela Current, flows northward toward the Equator along the southwest coast of Africa. On the north and south sides of the gyre are the westward-flowing South Equatorial Current and the eastward-directed South Atlantic West Wind Drift, respectively.

The system of currents in the South Pacific and the South Indian Ocean is somewhat more complicated than the above-mentioned gyres due to the presence of the two land masses, Australia and New Zealand. The cold, northward-flowing current on the eastern side of the gyre in the South Pacific Ocean is the Peru Current. On the western side of the gyre, in the South Indian Ocean, a warm, southward-moving current, known as the Mozambique Current, flows along the east coast of Madagascar and Africa. Farther south, the Agulhas Current washes the southeast coast of Africa.

In addition to the currents associated with subtropical gyres, there are at least four other major currents. Three of these are cold currents, flowing equatorward from the Arctic or Antarctic along the northeast coasts of the continents. In the North Atlantic we find the Labrador Current, in the South Atlantic the Falkland Current, and in the North Pacific the Oyashio Current. The fourth major current is the eastward-flowing Equatorial Countercurrent, which lies embedded between the North and South Equatorial Currents, and which is somewhat more conspicuous in the Pacific than in the Atlantic.

The currents are, in general, driven by the winds. The gyres reflect the influence of the large, whirling wind patterns called the subtropical anticyclones. On the equatorial side of these wind gyres are found the westward-blowing trade winds, which maintain the North and South Equatorial Currents. On the poleward side of the wind gyres are found the prevailing westerlies (winds blowing from west to east), which drive the Atlantic and Pacific currents (the west-wind drifts) from west to east.

These currents play an important role in modifying the weather and climate of the world. In the North Atlantic, the flow of relatively warm water around Iceland and the Scandinavian Peninsula keeps the Arctic ports free of ice even in winter, and maintains a relatively moderate air temperature in what would otherwise be a much colder region. The Kuroshio (Japan) Current plays a similar role in the Aleutian Island region. Over the cold currents, such as the Labrador and California Currents, the cooling of warm, humid air produces dense sea fogs.

The major surface currents of the ocean are driven by the stress of the wind on the sea surface. The wind tends to drag the surface water, by friction, along the wind direction. However, because of the rotation of the earth, the water does not move in the direction of the wind. Instead, the surface water moves in a direction to the right of the wind in the Northern Hemisphere and to the left of the wind in the Southern Hemisphere, as shown in Fig. 40. Thus a wind from the south (in the Northern Hemisphere) sets up a current flowing toward the northeast. The wind-driven current in the sea diminishes in speed with increasing depth, and also turns more to the right in the Northern Hemisphere (to the left in the Southern Hemisphere). The variation of the wind-driven current with depth is described by a theoretical spiral, called the Ekman spiral after the oceanographer who developed this concept. At a depth not more than 100 to 200 meters below the sea surface, these *shallow* wind-driven currents, known also as *wind-drift* currents, vanish.

Figure 40

One result of the wind stress on the sea surface is the phenomenon known as *upwelling*. This is a term used to describe the rising of cold water from deeper layers to replace warm surface water. One of the many places where upwelling commonly occurs is along the coast of California (San Francisco area). Here, the wind blowing from the north sets up the

California current, and at the same time transports warm surface water away from the coast. This shifting allows the colder bottom water to rise to the surface. The result is cold surface water, which is rather uncomfortable for most swimmers. The upwelling increases the abundance of plankton (tiny marine organisms on which fish feed) in the sea, and thereby improves fishing.

Another effect of wind, particularly strong winds such as those in hurricanes, is the *storm surge*. A strong wind blowing on the sea surface for many hours will pile up the water on the coast, often producing widespread death and destruction. Storm surges are responsible for the major part of hurricane devastation. Low-lying areas, such as the Gulf Coast of the United States, are especially vulnerable to the destructive effects of storm surges. In the 1900 hurricane in Galveston, Texas, 6000 people lost their lives when a storm surge inundated the city. (Storm surges are sometimes erroneously referred to as tidal waves.)

In certain bays and estuaries, strong currents are produced by the tidal ebb and flow of the sea. When the water is forced to pass through a narrow inlet, the tidal currents may be very strong. One of the most famous great tides of the world is the one in the Bay of Fundy, between Maine and Nova Scotia, where the range from high to low water may be as much as 15 meters. However, the tidal current is not excessively strong at the opening of this bay due to its great width. In narrower openings, on the other hand, the tidal currents may reach speeds as high as 5 meters per second (10 knots).

The currents in the sea are associated with differences of water pressure that exist from one place to another. These pressure differences are the result of density variations in the sea and of the slope of the sea surface.

The density of seawater varies with temperature and salinity. Warm freshwater is light, whereas cold, saline water is heavy. At great depths in the sea (e.g., below 1500 meters), the density is nearly uniform. The principal density variations occur closer to the sea surface. Where the density of the water is low, the water occupies a relatively large volume and bulges upward. Where the density of the water is high, the water is compressed, occupies a small volume, and shrinks downward, as shown in Fig. 41(a). At a certain depth below the sea surface, the thin, cold column of water weighs the same as the thick, warm column, and there is no variation of pressure in the horizontal. Above this level, the sea surface slopes downward from low to high water density. Because of the slope of the sea surface, the water pressure in the upper layers of the sea is higher where the density is low, and this pressure difference generates a force which tends to push the surface water toward the region of high density. In other words, the water tends to "roll downhill," i.e., down the slope.

However, the water does not actually slide downhill, due to the earth's rotation which deflects its downhill course. As we saw from Foucault's

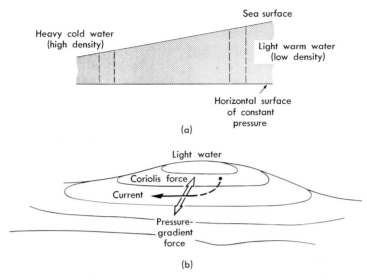

Figure 41

pendulum experiment, the earth's surface revolves in a counterclockwise sense in the Northern Hemisphere and clockwise in the Southern Hemisphere. The rotation gives rise to an apparent deflection of the motion of bodies on the earth, i.e., to the observer located on the rotating earth, it seems as if projectiles, winds, ocean currents, etc. were acted upon by a deflecting force which causes them to behave as if they were being pushed to the right (in northern latitudes) or to the left (in southern latitudes). This apparent deflecting force of the earth's rotation is called the *Coriolis force,* after a French mathematician of the eighteenth century. The Coriolis force acting on an ocean current deflects the course of the water until it is flowing parallel to the slope, e.g., into the page in Fig. 41(a). In the Northern Hemisphere, the cold high-density water will lie to the left of the current if we face downstream. (The reverse is true in the Southern Hemisphere.) The force due to the pressure difference (called the *pressure-gradient force*) is then balanced by the Coriolis force, and the current is said to be *geostrophic.* The balance of forces associated with the ocean currents is illustrated in Fig. 41(b).

The *deep* wind-driven currents, such as the Gulf Stream, are geostrophic in character. In these currents the water masses tend to arrange themselves with the lighter water on the right of the current and the heavier water on the left (in the Northern Hemisphere). Thus in the Gulf Stream, for example, the cold, heavy water is on the left, close to the coast, whereas the warm, light water is on the right, farther out to sea. The gradient of temperature and density across the current is largest where the current is strongest.

The speed of the current decreases with increasing depth, and disappears at some great depth where the density variations vanish. This level is known as the "level of no motion"; below it the current may reverse direction. (The "level of no motion" in the Gulf Stream varies from about 500 meters near the edge of the continental shelf to about 2000 meters on the deep, right-hand side of the current.) Oceanographers can calculate the geostrophic ocean currents relative to this "level of no motion" from the horizontal and vertical distribution of density in the sea. (The formula for the geostrophic ocean currents involves the latitude and the spatial variation of water density.) The density is determined from careful measurements of temperature and salinity. On oceanographic research cruises, profiles of density are constructed from these measurements, and the currents are then calculated from the density profiles. The computed currents agree, in general, with measured currents, but the agreement is not perfect.

It is now known that there are ocean currents in the deep water as well as near the surface. These deep-water currents flow rather slowly, but they play an important role in the worldwide exchange of water masses. Cold, heavy water sinks to great depths at high latitudes in the North Atlantic Ocean and, traveling two miles below the surface, spreads southward across the Equator, with the result that the deep water in the tropics is very cold. (Cool, intermediate water rises near the Equator, keeping the water temperature there lower than it would otherwise be.) At the same time, cold, heavy water also sinks in the Weddell Sea near Antarctica and flows northward as bottom water. It is these sinking cold-water masses which carry oxygen down to the ocean depths, making life possible even on the ocean floor. Kuenen refers to these sinking polar water masses as the "lungs of the ocean," because of their effect in ventilating the deep water.

The complexity of the subsurface currents is illustrated by the Cromwell Current or Pacific Equatorial Undercurrent. This is a narrow, fast-moving (about a meter per second) current, flowing eastward at a depth of 50 meters below the westward-flowing Pacific South Equatorial Current.

Ocean waves. The awesome destructive power of ocean waves is well known to anybody who has ever watched the sea or merely read about it. For many years, attempts have been made to predict the height and power of ocean waves. As a result of careful scientific studies considerable progress is now being made in understanding and predicting the phenomena associated with ocean waves.

In shallow water near coasts, the height of waves may be measured and recorded with a pressure-type wave recorder. This device, which is placed on the ocean bottom, records waves by "weighing" the mass of water above it as the waves move by. Another device used for observing and recording waves in shallow water is the wave staff, which may be used to record the height to which the water rises on the staff. In deeper water, a special type of sea anchor can be used to make the wave staff float at constant height above the deep, undisturbed water. The observer then notes the rise and fall of the sea surface relative to the staff.

For information about waves in deep water, oceanographers have in the past depended mainly on visual evidence supplied by shipboard observers. Recently, two types of automatic wave recorders have been developed for deep water. One of these, known colloquially as the "splashnik," is a floating accelerometer, which measures the vertical accelerations of the sea surface. A radio transmitter on the splashnik communicates the wave information to the oceanographic research vessel. The second type of device is a ship-borne wave recorder, consisting of a pressure recorder which measures the height of the waves relative to the ship, and a ship-motion recorder which measures the motion of the ship relative to the earth. A combination of these two measurements provides a record of the height of the waves encountered by the vessel.

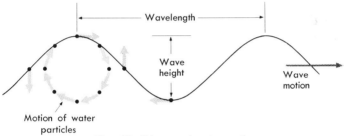

Fig. 42. Wave and water motion.

Waves are described and classified in terms of the following properties: length, period, frequency, wave speed, and amplitude or height. The wavelength of a wave is the distance from any phase of a wave to the next repetition of that phase, e.g., the distance from crest to crest or from trough to trough (see Fig. 42). The period of a wave is the time required for one full wavelength to pass an observation point, e.g., the time interval between the passage of two successive crests. The frequency of a wave is the reciprocal of the period, i.e., the number of waves passing an observation point per unit of time. The speed of the wave (also called wave speed, phase speed, or speed of propagation) is the rate at which a phase of the wave (e.g., a wave crest) travels. The wave speed is equal to the wavelength divided by the wave period. The height of a wave is the vertical distance from the top of the wave crest to the bottom of the wave trough. (This is the distance to which we refer when we speak, for example, of a 20-meter wave.) The amplitude of a wave is one-half the wave height.

There are two distinct classes of ocean surface waves: those in which the depth of the water is relatively shallow compared with the wavelength, and those in which the water is very deep compared with the wavelength. The former are called long waves or shallow-water waves. The latter are called short waves or deep-water waves. Long waves move with a speed* that depends only on the depth of the water, and hence travel through a given body of water with the same speed. Short waves on the other hand, move with a speed* that depends only on the wavelength of the waves, and hence travel with different speeds (the longer ones travel faster than the shorter ones), with the result that they spread out or overtake one another and interfere with one another. (Many waves are intermediate between long and short waves.) Both are referred to as *gravity waves*, because gravity is the "restoring force" that maintains the waves.

When a wave moves toward a beach from the open ocean, it changes from a deep-water to a shallow-water wave. As the depth of the water decreases toward the beach, the wave speed of the approaching wave also decreases, and this gradual decrease in turn causes the wavelength to decrease. This change in the character of a wave moving into shallow water is described by saying that the wave is "feeling the bottom." The effect of shallow water on the wave is to steepen the wave, ultimately causing it to break (see Fig. 43).

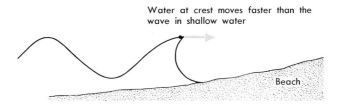

Fig. 43. Breaking of waves on a beach.

The waves one sees in the open ocean are progressive waves, i.e., the waveform moves. However, in lakes and some bays, so-called *standing* or *stationary waves* may develop. These are known as *seiches*. In a standing wave there are nodal lines (nodes) along which there is no movement of the water surface. On both sides of the node, the water surface rises and falls periodically, but the waveform does not move horizontally. The period of oscillation of these seiches depends on the length of the basin, the depth of the water, and the number of nodal lines. (See Laboratory Exercise 5.)

With a few exceptions, all waves are generated by the wind driving against the water. (So-called tidal waves or *tsunamis* are created by sub-

---

* The formula for the speed of long waves in shallow water is $c^2 = gH$, where c is the wave speed, g is the acceleration due to gravity, and H is the depth of the water. The formula for the speed of short waves in deep water is $c^2 = gL/2\pi$, where L is the wavelength.

marine earthquakes or volcanoes. Waves may also be generated locally by calving of icebergs or by rock slides.) The exact manner in which the wind develops waves is not completely understood. However, it is known that the height of the waves depends on the speed of the wind and on the *fetch,* the distance across which the wind has been blowing over the water. Waves are generated in storms at sea and then travel outward from the generating area, often giving advance warning of the approach of the storm. As the waves travel away from the generating area, the shorter waves die out, leaving only the very long waves, called *swell,* to travel over great distances. The long swell is distinguished from the shorter, more irregular waves of all lengths which are found close to the storm area itself, and which are referred to as *sea.*

We must distinguish between the motion of the *wave* and the motion of the water *particles.* While the wave is moving across the sea surface, the water particles execute circular orbits, as shown in Fig. 42. Consider a water particle on a wave crest. Here the water is moving in the direction of the wave. As the next trough approaches, the water level between crest and trough must be sinking. Thus, after the passage of the crest, our water particle begins to sink. The sinking of the water surface requires that the water masses at the trough and crest separate and move in opposite directions. Thus the water at the trough is moving in a direction opposite to the direction in which the wave is moving. When the trough passes, our water particle starts moving in a direction opposite to that in which the wave is moving. With the passage of the trough, the water surface rises locally, and our water particle goes up. With the passage of the next crest, the water particle begins another circular orbit while remaining in the same location. The motion of the water particles, as contrasted with the wave motion, can be seen by observing the motion of floating objects as the waves pass under them. A wave breaks when the water on the crest advances faster than the wave itself. This occurs in shallow water, where the wave, feeling the bottom, is retarded, while the water at the crest continues its rapid advance (Fig. 43).

Tides. The regular rise and fall of the sea, known as the *tide,* can be observed on any shore. Tides have been observed, measured, and recorded (in memory, if not on paper) for as long as man has lived near the sea. The association between the tides and the moon has been known almost as long as the periodic changes of the tides themselves.

The theory of tides and methods for analyzing the tides into their periodic components were developed in the eighteenth and nineteenth centuries, with some improvements added in the twentieth century.

Tide gages, which are similar in principle to wave recorders, are employed to record the height of the sea surface on a clock-driven chart, thus showing the variation of water level with time. From an examination of the tide record, it is immediately apparent that there are two high tides and two low tides daily. For example, on four days in March 1960, the following times of high tide were recorded in New York Harbor.

Date	Time of high tide	
	a.m.	p.m.
3/28	9.05	9.25
3/29	9.48	10.08
3/30	10.32	10.51
3/31	11.16	11.33

We note that the period of the tide, i.e., the time interval between high tides, is about 12 hours 20 minutes, so that high tide occurs a little later each day. If the tides were controlled by the sun, they would occur at the same time every day. However, the tides are controlled primarily by the moon. The length of a lunar day (i.e., the interval between successive lunar transits of the meridian) is 24 hours 50 minutes, and the half-lunar day is 12 hours 25 minutes long. The tidal period corresponds rather closely to the length of the half-lunar day. Thus the tide rises and falls by lunar time rather than by solar time.

The tides are caused principally by the gravitational attraction between the earth and the moon. (There is also a tidal effect of the sun which we shall discuss shortly.) The mechanism of the tide is illustrated in Fig. 44. The gravitational pull of the moon is greater on the side of the earth facing the moon than it is on the opposite side, in accordance with Newton's inverse-square law. The force is at a maximum at the point nearest to the moon, and at a minimum at the point farthest from the moon on the opposite side of the earth. The lunar gravitational force causes the water on one side of the earth to be pulled upward *toward* the moon, whereas on the opposite side, where the gravitational pull of the moon is weaker, the water bulges upward *away* from the moon.* Thus two equal bulges develop on both sides of the earth. As the earth rotates relative to the moon, the water level falls and rises with a period of half a lunar day, so that two maxima and two minima in the water level occur each lunar day.

The two high tides are generally of different magnitude, due to the *declination* of the moon (its elevation above the equatorial plane), as illustrated in Fig. 45. The two maxima in the water level occur on a line passing from the center of the moon through the center of the earth. As the earth rotates, a given point on a latitude circle does not, in general,

---

* The gravitational forces are balanced by equal and opposite centrifugal forces, which are due to the rotation of the earth-moon system about its common center of mass.

CHAP. 3] CURRENTS, WAVES, AND TIDES IN THE OCEAN 97

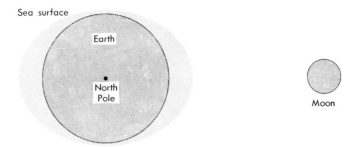

Fig. 44. The tide-producing effect of the moon (exaggerated).

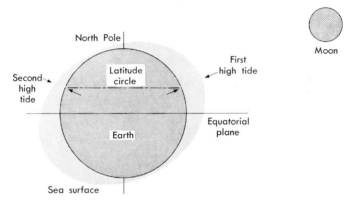

Fig. 45. The diurnal inequality of the tides.

pass through equally high water levels on the two sides of the earth unless the declination of the moon is zero. Hence one of the daily high tides is generally higher than the other.

The sun also exerts a tidal influence on the ocean. It can be shown that the tidal force of an extraterrestrial body is proportional to its mass and inversely proportional to the cube of its distance from the earth. Although the mass of the sun is far greater than that of the moon, its distance from the earth is also far greater so that its tidal influence is only about half as large as that of the moon. Nevertheless, the effect of the sun is significant, although not dominant. When the sun and moon are lined up with the earth, the tidal effects of the two bodies reinforce each other, and the tides are at their maximum height. The tides, called *spring tides* (although they have nothing to do with the season of the year), occur when the moon is in conjunction or opposition, i.e., at the times of new and full moon. When the gravitational forces exerted by the sun and the moon on the earth are pulling at right angles to each other, they do not reinforce each other, and the tides are weakest. These tides, called *neap tides,* occur at quadrature (half-moon). Thus the tides follow the phases of the moon.

There are additional effects due to the variable distances between the moon and earth, and the sun and earth, as well as the more important effect of varying lunar and solar declination. As a result, the tide is a quite complicated phenomenon. To forecast the tides it is necessary to analyze the observed tide into its periodic components. The principal components are the lunar semidiurnal (half-lunar day) tide, the solar semidiurnal tide, and a combination of the lunar and solar diurnal (24-hour) tides. The amplitude of each component is computed by *harmonic analysis,* a technique which involves the use of a special computer called a *tide analyzer.* Then each of the known components, with its known period, is predicted for as far in advance as we like. By adding together the various predicted components of the tide (on a tide-predicting computer), we can predict future tides. Such tide forecasts, of course, do not take into account the effects of storms on the water level.*

REFERENCES

1. P. H. KUENEN, *Realms of Water.* New York: Wiley, 1955.
2. P. A. GORDIENKO, "The Arctic Ocean," *Scientific American,* **204,** 88–102 (May 1961).
3. G. E. R. DEACON, "The Oceans," in *The Earth and Its Atmosphere,* D. R. Bates, ed. New York: Basic Books, 1957.
4. E. T. EADY, "The General Circulation of the Atmosphere and the Ocean," in *The Earth and Its Atmosphere,* D. R. Bates, ed. New York: Basic Books, 1957.
5. R. L. FISCHER and R. REVELLE, "The Trenches of the Pacific," in *The Planet Earth,* A Scientific American Book. New York: Simon and Schuster, 1957.
6. W. H. MUNK, "The Circulation of the Oceans," in *The Planet Earth,* A Scientific American Book. New York: Simon and Schuster, 1957.
7. H. U. SVERDRUP, M. W. JOHNSON, and R. H. FLEMING, *The Oceans: Their Physics, Chemistry, and General Biology.* Englewood Cliffs, N.J.: Prentice-Hall, 1942.
8. W. S. VON ARX, *An Introduction to Physical Oceanography.* Reading, Mass.: Addison-Wesley, 1961.
9. J. A. KNAUSS, "The Cromwell Current," *Scientific American,* **204,** 105–116 (April 1961).
10. A. DEFANT, *Ebb and Flow: The Tide of Earth, Air, and Water.* Ann Arbor, Mich.: University of Michigan Press, 1958.
11. K. S. DAVIS and J. A. DAY, *Water, The Mirror of Science.* Garden City, New York: Doubleday, 1961.

---

* See references 9, 10 and 1 through 8.

IV. The Atmosphere

In this section we will study the atmosphere, the thin but vital gaseous envelope of the earth.

Meteorology is the science of the atmosphere. The word "meteorology" was used by Aristotle as the title of his treatise on the air more than 2300 years ago. It refers to the study of things "high in the air," including the air itself. Among the things in the air that concern the meteorologist is the composition of this mixture of gases we call air. Of special interest are the variable constituents of air, of which water vapor is the most important. The condensation of water vapor produces clouds and is largely responsible for our weather. The analysis and prediction of weather are two of the better-known practical activities of meteorology.

Man looks up into the ocean of air from below, like a crab on the bottom of the sea. What is the vertical structure of this atmospheric envelope? What is "up there"? For a long time the upper atmosphere was unknown, unexplored. Today rockets and meteorological satellites enable the meteorologist to probe the atmosphere to unlimited heights and to survey it from above. We examine the vertical structure of our atmosphere in Section 2.

The air is a gas in motion. The study of winds, the atmospheric currents which produce our weather and climate and cleanse the atmosphere of pollution, constitutes another large area of meteorology.

The atmosphere is irradiated by the sun. Solar radiation provides the energy that heats the atmosphere and drives the winds of the earth. The distribution of solar radiation is one of the principal factors influencing world climate.

There is now, in this age of rockets and satellites, greater interest than ever in the upper reaches of the atmosphere. What is the structure of the upper atmosphere? Of what is it composed? How is it heated? What are its electrical properties? How does it react to disturbances on the sun? And what is the relation between events in the upper and in the lower atmosphere? We shall try to answer some of these questions below.*

1. Air. *Chemical composition.* The earth is wrapped in an envelope of tenuous gas known as the atmosphere. The ancient Greeks considered this gas, which they called air, one of the four basic elements, along with fire, water, and earth. In the eighteenth century the great "pneumatic chemists," Black, Scheele, Priestley, Lavoisier, Cavendish, and Rutherford, showed that air is in fact a mixture of several gases. Subsequent

---

* References 1 through 10 apply to all sections of this chapter.

investigations have shown that the chemical composition of air is remarkably uniform.

The air contains certain permanent and some variable constituents. Of the latter, the most important in the lower atmosphere is water vapor, which may, in very humid air, constitute as much as two or three percent of the gas, but usually comprises less than one percent. Of the remaining "dry air," 78 percent (by volume) is nitrogen (N_2) and 21 percent (by volume) is oxygen (O_2). The remaining one percent is made up of argon (A), carbon dioxide (CO_2), and many other gases in minute quantities.

The first to recognize that air consists of two parts, an inert gas (now called nitrogen) and one that supports life and combustion (now called oxygen), was the seventeenth-century chemist John Mayow. Nitrogen was identified by Rutherford in 1772, and oxygen was isolated by Scheele in 1773 and, independently, by Priestley in 1774. The role of oxygen in combustion and life was shown by Lavoisier in 1774. Cavendish discovered argon in 1785. Carbon dioxide had already been described by Black in 1755.

The chemical composition of dry air is nearly the same all over the world. Even up to an altitude of 80 kilometers (50 miles) there is almost no change in the composition of the air, indicating that the air is well mixed. If it were not, the gases would separate out, with the heavier gases at the bottom and the lighter ones at the top.

One of the important variable components of the air is the triatomic form of oxygen called *ozone* (O_3) which occurs in *relatively* high concentration (up to 10 parts ozone per million parts air) in an atmospheric layer extending from about 20 to 40 kilometers above the earth. The importance of this minor constituent of the upper atmosphere is due to the fact that it absorbs much of the sun's ultraviolet radiation that would otherwise produce lethal effects on earth.

At very high levels in the atmosphere (above 90 kilometers), the oxygen is found in atomic (O) rather than molecular (O_2) form.

Radioactive gases. The disintegration of radioactive elements in rocks has led to the presence of very small quantities of radioactive gases in the atmosphere. The principal radioactive gases are radon (from radium), thoron (from thorium), and actinon (from actinium). All have very short half-lives. Since they are emitted by rocks, they are found in much greater abundance over land than over water.

Radioactive carbon (the isotope of carbon called carbon-14) is produced in the upper atmosphere by the bombardment of nitrogen with neutrons produced by cosmic rays. This isotope of carbon, with its half-life of 5760 years, combines with oxygen to produce radioactive carbon dioxide which enters living matter. From the ratio of carbon-14 to carbon-12 (ordinary carbon) in dead matter, it is possible to determine the time of death. This technique is known as *radiocarbon dating*.

Since 1945, man has been adding significant quantities of artificial radioactivity to the atmosphere, principally through the explosion of atomic bombs.

Aerosols. Suspended in, or slowly falling through, the air is a myriad of tiny solid and liquid particles. Among these so-called *aerosols* are the water droplets and ice crystals in clouds, dust particles, microorganisms (spores, pollens, molds, etc.), and condensation nuclei. Condensation nuclei, the most numerous of the aerosols, are invisible particles, or droplets, which have an affinity for water. When cloud or fog droplets form by condensation in the atmosphere, the water first condenses on these nuclei. They are found in greatest abundance over cities, and decrease markedly with altitude. Sulfates and salt particles are the main condensation nuclei.

Atmospheric pollution. Human activities, such as the burning of sulfur-bearing fossil fuels (coal and oil) and the operation of gasoline-burning automobiles, lead to the discharge of great quantities of toxic gases, particles, and smoke into the atmosphere. At times, the concentrations of these pollutants exceed tolerable limits, as, for example, in the case of Los Angeles smog. One of the pollutants is sulfur dioxide, which converts to sulfur trioxide under the action of sunlight and then combines with water to produce dilute sulfuric acid. Among the more important irritants in cities are certain oxidants (including ozone) produced by the action of sunlight on the exhaust fumes of automobiles.

Water vapor. The principal variable component of air is water vapor. This invisible gas is lighter than oxygen and nitrogen. Hence humid air, under the same conditions of pressure and temperature, is less dense (lighter) than dry air. The quantity of water vapor in the air, usually referred to as *humidity,* can be measured and expressed in various ways, some of which are described below.

Absolute humidity indicates the mass of water vapor in a unit volume of air, i.e., the density of water vapor. (Meteorologists rarely use, and almost never measure, this quantity.)

Vapor pressure is that part of the total air pressure which is exerted by the water vapor, i.e., the partial pressure of the water vapor. (This is generally determined indirectly from other measurements, if at all.)

Relative humidity is the ratio of the actual vapor pressure to saturation vapor pressure, the latter being the vapor pressure which exists when the vapor is in equilibrium with liquid water. (Sometimes this saturation pressure is erroneously called maximum vapor pressure.) Relative humidity is expressed in percent, and can be measured with a hair hygrometer, for hair (human or horse) stretches with increasing relative humidity. The vapor is said to be saturated when the relative humidity is 100 percent.

Dew point is the temperature to which vapor must be cooled (at constant pressure) to produce saturation. This saturation temperature can be determined with a dew-point hygrometer, a device which cools the air until dew appears on a metal surface. When the relative humidity is 100 percent, i.e., at a condition of saturation, the dew point is equal to the temperature. Normally the dew point is lower than the temperature.

Wet-bulb temperature. One of the standard humidity instruments is the *psychrometer* (cooling meter), consisting of a dry-bulb thermometer and a wet-bulb thermometer, the latter being moistened with wet muslin wrapped around the bulb. Evaporation lowers the temperature of the wet bulb. The drier the air, the greater the evaporation and cooling. The difference between wet-bulb and dry-bulb temperatures can be used to determine humidity. The human body also acts as a wet-bulb psychrometer. We feel cooler in hot dry weather than in hot humid weather because of the evaporative cooling of surface perspiration.

2. Vertical structure of the atmosphere. *Observations of the upper atmosphere.* The earliest efforts to measure the change of air temperature with height above the earth were made in the eighteenth century, first with kites and then with manned, gas-filled balloons. At the end of the nineteenth century, self-recording meteorological instruments for measuring pressure, temperature, and humidity were carried aloft to great heights in the atmosphere on small unmanned balloons. The records recovered from these *ballons-sondes* made it possible to determine the vertical structure of the air and led to the discovery of the stratosphere (around 1900).

A great acceleration in upper-air measurements came with the invention (about 1930) of the *radiosonde*. This meteorological sounding device measures pressure, temperature, and humidity, and transmits the information to earth immediately by radio, thus making recovery of the instrument unnecessary. Since radiosondes are now used for routine measurements every day at hundreds of observing stations throughout the world, the meteorologist receives a constant flow of information enabling him to construct up-to-date maps of the upper atmosphere at daily or even more frequent intervals.

The altitude reached by the radiosonde balloon rarely exceeds 30 kilometers. To reach higher levels of the atmosphere, meteorologists now use rockets. A rocketsonde carries the meteorological instrument package to a selected altitude and ejects it. The miniature weather station then records pressure, temperature, and humidity as it floats to the earth on a parachute.

Before the advent of rockets, knowledge about the vertical structure of the atmosphere above 30 kilometers was gleaned indirectly from the study of meteor trails, the refraction of sound waves, and theoretical calculations. Rocket measurements have verified our expectation about the general vertical structure of the atmosphere.

The layers of the atmosphere. The vertical structure of the atmosphere may be described in terms of the change of temperature with height. As

CHAP. 4] VERTICAL STRUCTURE OF THE ATMOSPHERE 103

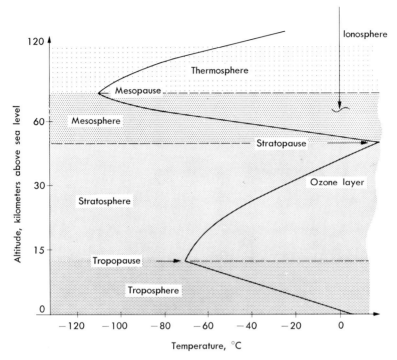

Fig. 46. Variation of temperature with height in the atmosphere; the layers of the atmosphere.

shown in Fig. 46, the temperature generally decreases with increasing altitude above the surface of the earth. This trend exists because the lower atmosphere is heated from below. Clear air is nearly transparent to most of the radiation from the sun, which therefore passes almost unabsorbed through the air to the ground. The ground is heated by the insolation which it absorbs. The heated earth radiates long-wave, invisible, infrared radiation, which the air (i.e., the carbon dioxide and water vapor in the air) does absorb. Thus, since the air is heated from the bottom, the temperature diminishes with increasing height. This decrease of temperature with height is called the *lapse rate*.

Layers of air in which the temperature falls with increasing altitude are subject to active vertical mixing and overturning. Layers of air in which the temperature does not change, or perhaps even rises, with increasing altitude are inert, sluggish, resist vertical mixing, and are said to be *stable*. An increase of temperature with height is called an *inversion*.

The lowest layer of the atmosphere (Fig. 46) is called the *troposphere* (Gr. *tropos,* turning), which is characterized by strong vertical mixing, overturning, turbulence, and weather. The average depth of the troposphere is about 12 kilometers (8 miles), and it contains more than 80 percent of the mass of the atmosphere. The upper limit of the troposphere, called the *tropopause,* is higher in the tropics (about 16 kilometers) and lower in the polar regions (as low as 8 kilometers). At the tropopause the temperature falls to as low a value as $-80°C$ ($-112°F$) over the tropics. The tropopause is warmer in higher latitudes. For example, in the polar regions the temperature at the tropopause is about $-55°C$ ($-67°F$) on the average.

Above the tropopause is the second layer of the atmosphere, known as the *stratosphere.* The term refers to the fact that this layer is stratified, and that vertical mixing is weak. This stratification and absence of violent overturning are due to the increase in temperature with increasing altitude in the stratosphere. (In some regions of the stratosphere, the lapse rate is isothermal, i.e., the temperature is constant with height.) The stratosphere extends from the tropopause to an altitude of about 50 kilometers (30 miles).

The upper boundary of the stratosphere is known as the *stratopause.* Unlike the tropopause, which is a layer of minimum temperature, the stratopause is a layer of maximum temperature; here the temperature may be higher than that at the ground. The stratopause coincides with the top of the ozone layer, and its high temperature is due to the absorption of ultraviolet sunlight by ozone. The ozone layer acts as a second heat source (in addition to the earth) for the atmosphere. Since the temperature decreases as the distance from a heat source increases, the temperature decreases downward from the ozone layer and upward from the earth, producing a temperature minimum in between—at the tropopause.

Above the stratopause, the temperature again decreases with increasing altitude up to a height of about 80 kilometers (50 miles). This layer of the atmosphere extending from 50 to 80 kilometers above sea level is called the *mesosphere* (Gr. *meso,* middle). Like the troposphere, the mesosphere is probably a layer of considerable turbulence and overturning. The top of the mesosphere is called the *mesopause.* This layer exhibits the lowest temperature in the atmosphere, about $-110°C$ ($-166°F$).

Above the mesopause the temperature rises rapidly with height in the region of the atmosphere called the *thermosphere.* The high temperature of the thermosphere is a result of absorption of solar radiation by molecular oxygen and, with increasing altitude, by molecular nitrogen and by helium. At very high levels direct heat conduction from the sun is effective. The thermosphere also encompasses most of the region of the atmosphere known as the *ionosphere.* In the ionosphere, relatively high concentrations of electrons and ions (electrically charged particles) are present in the form of conducting layers which reflect radio waves back to earth. Although the lowest layers of the ionosphere (e.g., the D-layer) lie below the mesopause, the principal layers (the E- and F-layers) are in the thermosphere.

Starting at about 600 kilometers above the earth, the "air" consists of isolated atoms or molecules between which interactions are relatively weak. This is the beginning of the region known as the *exosphere,* the outer fringe of the atmosphere, including the Van Allen radiation belts. (It should be noted that the density of air falls off quite rapidly with height, so that even at as low an altitude as 50 kilometers above sea level, the top of the stratosphere, the air density is less than one-thousandth the density of air at sea level.)

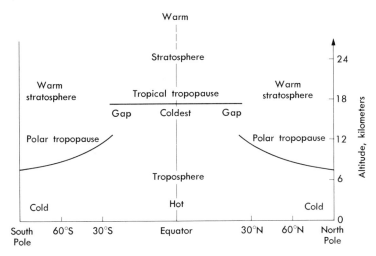

Fig. 47. Pole-to-pole vertical cross section through the atmosphere.

Vertical cross section through the atmosphere from pole to pole. The structure of the atmosphere is shown in Fig. 47 in a vertical cross section through the atmosphere from pole to pole. The cross section shows the variation in tropopause structure with latitude. Note the high, cold tropical tropopause and the low, warm polar tropopause. The two tropopauses overlap in the subtropics, where there is a gap, or break, in the tropopause.

In the troposphere, the temperature falls from the Equator toward the poles. In the lower stratosphere, the temperature is lowest over the Equator and increases toward the poles. We shall see later that this reversal of the north-south temperature gradient from troposphere to stratosphere causes the westerly winds to reach their maximum speed at the tropopause, and to decrease with increasing height in the lower stratosphere.

Pressure and altitude. Just as one may determine depth in the ocean by measuring the pressure below the sea surface, so one may determine altitude above sea level by measuring the air pressure. This is the principle of the altimeter, a barometer (pressure meter), used by airplane pilots and mountain climbers to determine their altitude.

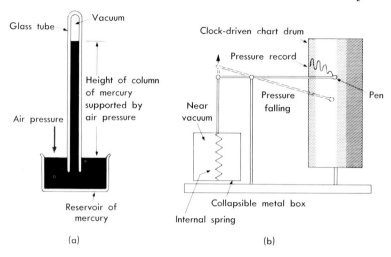

Fig. 48. (a) Mercury barometer (after Torricelli). (b) Aneroid barograph.

In 1643, Evangelista Torricelli, successor to Galileo, invented the mercury barometer (Fig. 48a) to demonstrate the weight of air. Atmospheric pressure is the weight of air per unit horizontal area. Torricelli showed that at sea level the weight of the air can support a column of mercury about 76 centimeters (30 inches) high, and concluded that the total column of air from the ground up weighs the same (per unit area) as a 76-centimeters column of mercury.

Shortly after the invention of the barometer, the philosopher Blaise Pascal predicted that the height of the column of mercury in the barometer should be smaller on top of a mountain than at sea level because the pressure and weight of the air are proportional to the mass of air *above* the observer. Pascal's prediction was verified by his brother-in-law, who undertook to carry a barometer up a mountain.

The pressure of the mercury in a barometer (and, therefore, the atmospheric pressure) is equal to the product of the height of the mercury column times the density of the mercury times the acceleration due to gravity. Meteorologists express the pressure in units of *millibars* (abbreviated mb). (A millibar is equal to 1000 dynes per square centimeter, the dyne being the unit of force in the cgs system.*) The pressure at sea level is *about* 1000 millibars (or about 14.5 pounds per square inch) and is variable.

The change of pressure with altitude depends on the density of the air. Air density is more variable than water density, so that measuring altitude with a barometer (altimeter) in the atmosphere is not so simple as meas-

---

* In the "centimeter-gram-second" (abbreviated cgs) system, mass is expressed in grams, length in centimeters, and time in seconds.

Table 8
Average Pressure, Expressed in Percent of Sea Level Pressure, Versus Altitude Above Sea Level

Altitude kilometers (miles)			Pressure, % (approximate)
0	(0)		100
1.5	(1)		85
3	(2)		70
6	(3.5)		50
10	(6)		30
12	(8)	Average tropopause	20
16	(10)	Tropical tropopause	10
30	(20)		1
50	(30)	Stratopause	0.1
80	(50)	Mesopause	0.001
160	(100)		0.0000005

uring depth with a barometer (bathymeter) in the ocean. To determine altitude *accurately* from the pressure, we must also know the air temperature and the humidity. However, we can determine altitude *approximately* from a measurement of atmospheric pressure.

Mercury barometers are used only for precise measurements of pressure in meteorological stations. For altimeters and for recording barometers, called barographs, *aneroid* barometers are used. These nonliquid barometers consist of one or more evacuated metal boxes, supported by springs, which expand when the outside air pressure falls and collapse when the outside air pressure rises (see Fig. 48b).

The pressure at any level in the atmosphere is proportional to the mass of air above that level. At an altitude of 6 kilometers above sea level, for example, the pressure is about 500 millibars, or one-half the pressure at sea level. Thus half the mass of the atmosphere is below 6 kilometers (about $3\frac{1}{2}$ miles), and half is above that altitude. Table 8 shows the average (approximate) pressure at various altitudes above sea level expressed in percent of the sea-level pressure. These numbers also represent the fraction of the total mass of the atmosphere above any level in the atmosphere.

Note in Table 8 that 90 percent of the atmosphere lies below the tropical tropopause (16 kilometers above sea level), 99 percent lies below an altitude of 30 kilometers, 99.9 percent lies below the 50-kilometer-high stratopause, 99.999 percent lies below the 80-kilometer-high mesopause, etc. These percentages may help to answer the question, How high is the atmosphere? Note also that the depth of the atmospheric layer containing

99.99 percent of the atmosphere is only about one one-hundredth (0.01) as large as the radius of the earth. The atmosphere is indeed, at least in this sense, a very thin gaseous shell surrounding the earth.

3. Wind. From the smallest-scale air motions in dust whirls to the general global circulation of the whole atmosphere, all winds are governed by certain basic laws of motion. The energy for these atmospheric motions is derived from the sun's radiation, via the nonuniform heating of the earth. The resulting pressure differences (gradients) produce forces which drive the air toward areas of low pressure. But the rotation of the earth then comes into play, causing apparent alterations in the wind direction. The result is a relation between wind and pressure that we can use to understand the general circulation as well as the circulation in smaller-scale wind systems such as hurricanes and other cyclones.

In this section, we shall discuss the laws of motion, the worldwide wind pattern, the circulation of air around high- and low-pressure areas, the variation of wind with height, and certain special air motions, such as tornadoes, sea breezes, and monsoons.

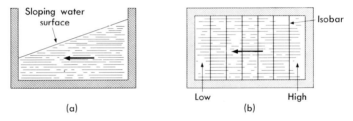

Fig. 49. Pressure gradient in a tub of water. (a) Vertical view. (b) Pressure distribution at the bottom of the tank as seen from above.

Wind and pressure. We may use a tub of water as an analogy to the atmosphere in order to understand how pressure gradients produce wind. Consider the water in the rectangular basin drawn in cross section in Fig. 49(a). We tilt the water surface as shown in the figure. (This may be done by pressing down on the water surface with a tilted board, and then removing the board.) The pressure at any depth is proportional to the depth, and therefore increases from left to right. Figure 49(b) shows the distribution of pressure at the bottom of the tank as it would been seen on a map. The lines on the map are lines of equal pressure, called *isobars.* The variation of pressure with horizontal distance, i.e., the *pressure gradient,* is indicated by the spacing of isobars, the variation being large where the isobars are close together.

The pressure is greater to the right than to the left of the figure. Therefore a net force is directed from right to left, i.e., from high to low pressure. This force drives the water across the isobars from high to low pressure (after the board is removed from the top of the water), and the water, as in a seiche, sloshes back and forth, driven by this *pressure-gradient force*. In the atmosphere, the pressure distribution is represented by isobars (lines of equal sea-level atmospheric pressure, for example) drawn on a weather map, and the pressure-gradient force points from high to low pressure.

The air would blow in the same direction as the pressure-gradient force if it were not for the rotation of the earth. In small-scale air motions, the effect of the earth's rotation is not important. But in larger wind systems, such as those seen on weather maps, it is important.

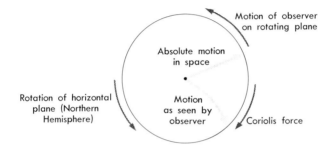

Fig. 50. The deflecting force of the earth's rotation (Coriolis force).

The Coriolis force, or deflecting force of the earth's rotation (see the section on currents, waves, and tides in the ocean, above), is illustrated in Fig. 50. The circle represents the horizontal plane tangent to the earth at any latitude. As shown by Foucault's pendulum, in the Northern Hemisphere this plane rotates in a counterclockwise sense about a vertical axis, and its angular velocity increases from zero at the Equator (no rotation) to a maximum (one revolution per day) at the pole. (In the Southern Hemisphere the rotation is clockwise.) The solid arrow in Fig. 50 represents an object (bullet, water, air) moving in a straight line in space. The observer watching this motion is on the rotating plane, and he moves away from the moving object. Therefore, to him the object appears to turn as if it were being pushed to the right of its course. The force which appears to be pushing to the right (in the Northern Hemisphere) as the observer looks downwind is called the Coriolis force, or deflecting force of the earth's rotation. (In the Southern Hemisphere it is directed toward the left.) Although the Coriolis force is an *inertial* force (like the centrifugal force), we may treat it as if it were a real force for the purpose of studying air motions on the earth.

The Coriolis force is proportional to the wind speed, and also varies with latitude, just as the angular velocity of the horizontal plane varies, being zero at the Equator. The force is perpendicular to the wind. The pressure-gradient force drives the air toward low-pressure areas, but the

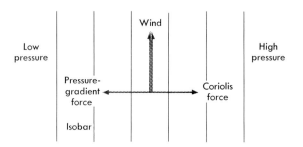

Fig. 51. Geostrophic wind (Northern Hemisphere).

Coriolis force causes the wind to turn until it is blowing nearly parallel to the isobars, as shown in Fig. 51. Now the two forces, pressure gradient and Coriolis force, are acting in opposite directions, and the wind may blow steadily under balanced forces. When the wind blows parallel to the isobars and the two forces exactly balance each other, the wind is said to be *geostrophic*.

From Fig. 51 we may deduce the law, known as Buys-Ballot's law (after a nineteenth-century Dutch meteorologist), concerning wind and pressure.

If an observer stands with his back to the wind, the pressure is lower (higher) on his left in the Northern (Southern) Hemisphere than on his right.

It can also be shown that the wind speed (for geostrophic wind) increases in proportion to the pressure gradient, so that the closer together the isobars, the faster the wind at any latitude.

Near the earth's surface, friction plays an important role in air motions, in that it causes the wind to slow down and to turn toward low-pressure areas. Thus, near the ground the wind obeys Buys-Ballot's law, but it does not blow exactly parallel to the isobars. Instead, it crosses the isobars in the direction of low pressure at an angle of about 20 to 45 degrees.

Cyclones and anticyclones. The word cyclone (Gr. *kyklos,* ring or circle) refers to the rotary air motion in storms. Today, *cyclone* means any region of low pressure as seen on a weather map, and includes large, weak low-pressure areas as well as the deep low-pressure centers in hurricanes. As shown in Fig. 52(a), it follows from Buys-Ballot's law that the circulation around *cyclones* (lows) must be *counterclockwise in the Northern Hemisphere.* Near the ground, where friction is important, the wind also tends to blow inward in the direction of low pressure, as shown in Fig. 52(b). The inflowing, or convergence, of the wind in cyclones causes the air to pile up, rise, and cool, resulting in clouds and precipitation. Lows, or cyclones, are therefore regions of poor weather.

The opposite of a cyclone is the *anticyclone,* or high-pressure center. As illustrated in Fig. 53(a), the wind in the Northern Hemisphere blows in a clockwise sense around anticyclones, or highs, in accordance with

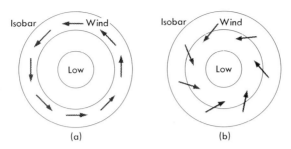

Fig. 52. Circulation around a cyclone in the Northern Hemisphere. (a) No friction. (b) With friction.

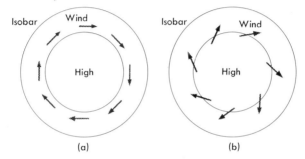

Fig. 53. Circulation around an anticyclone (high) in the Northern Hemisphere. (a) No friction. (b) With friction.

Buys-Ballot's law. Friction, as shown in Fig. 53(b), turns the wind outward from the high center. This outflowing, or divergence, of the wind in anticyclones causes the air to sink, warm, and dry out, so that highs, or anticyclones, are regions of fair weather.

In the Southern Hemisphere, the direction of the circulation is reversed. The wind spirals clockwise and inward in cyclones, counterclockwise and outward in anticyclones.

Variation of wind with height. Occasionally we may notice that the wind near the ground is blowing from one direction, but the clouds, driven by winds at higher altitudes, are moving from a different direction. What, then, are the "rules" governing the change in wind velocity with altitude?

The basic principles from which we can deduce the changes in wind with height are Buys-Ballot's law (and the geostrophic wind concept) and the *hydrostatic equation.* This equation states that pressure falls off more rapidly with altitude in dense (cold) air than in light (warm) air. We can combine these principles to deduce the change of wind with height.

Consider a cyclone in which the air is colder at the center than on the periphery. With increasing altitude, the pressure falls more rapidly over the center than over the periphery. As a result, the pressure gradient increases with increasing height, and the winds blowing around the cyclone become stronger. A little thought will show that a similar increase in circulation also takes place in an anticyclone whose center is *warmer* than the periphery. We see then that a combination of low temperature and low pressure on the one hand, or a combination of high temperature and high pressure on the other, causes the wind to increase with height.

Now consider the middle latitudes of our hemisphere, where the prevailing wind (as we shall see) is from the west. The pressure (according to Buys-Ballot's law) is lower to the north and higher to the south. The temperature is also lower to the north and higher to the south. Therefore, the pressure gradient and the west winds increase with increasing altitude.

In the stratosphere, the direction of the temperature gradient is reversed. The lowest temperature is found at the cold tropical tropopause, whereas the polar stratosphere at the same height is much warmer. Thus, in the lower stratosphere, we find warm air with low pressure, and cold air with high pressure. As we ascend, the pressure gradient becomes weaker in the stratosphere, and the west winds also diminish with height. The result is a maximum west wind at the tropopause. In the stratosphere, the west wind not only decreases with height but may actually reverse, so that at some altitude in the stratosphere the wind may be blowing from east to west.

The strong westerly winds are often concentrated in narrow "rivers" of very high wind speed (sometimes as high as 100 meters per second) which are usually found close to the gap in the tropopause. These high-speed currents in the upper troposphere are called *jet streams*. An aircraft flying from west to east will fly in the jet stream to increase its ground speed by taking advantage of the tail wind. But an aircraft flying from east to west will avoid this head wind.

Unlike the westerlies, which increase with height in troposphere, the tropical trade winds, blowing from east to west, decrease with increasing altitude. This reversal is due to the fact that, in the trade-wind belt, the pressure is low where it is warm, at low latitudes, and increases toward the colder higher latitudes. The temperature field causes this pressure gradient to disappear with height. At higher altitudes in the troposphere over the tropics, the wind blows from the west.

The planetary wind and pressure pattern. The *average* worldwide distribution of wind and pressure near sea level is shown in Fig. 54. Note that the winds blow in accordance with Buys-Ballot's law.

In constructing a picture of planetary circulation, we have averaged conditions over time and around the latitude circles, and have not included the influence of the continents (known as the *monsoonal effect*) which will be described later.

The low pressure region around the Equator is known as the equatorial low, the doldrum belt, or the intertropical convergence zone. Here the wind is light and variable, and showers are frequent due to the piling up of warm, moist air.

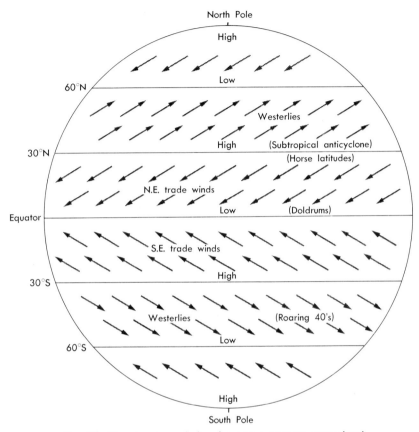

Fig. 54. The planetary wind and pressure patterns at sea level.

From the Equator, the pressure increases poleward in both hemispheres, reaching maximum values at about 30 degrees north and south. These are the subtropical high-pressure belts. Between the subtropical anticyclones and the doldrums are found the trade winds—very persistent winds blowing from the east. In the Northern Hemisphere, the trade winds blow from the northeast, and in the Southern Hemisphere, from the southeast.

In the subtropical high-pressure areas, known as the *horse latitudes,* the winds are light and variable so that sailing ships were often becalmed. This is also a region of low rainfall, sunny skies, deserts, and high ocean salinity.

On the poleward side of the subtropical high-pressure belt is the zone of westerlies. Here the wind is far more variable than in the trade region, due to the high frequency of traveling cyclones. But the prevailing wind direction is from the west.

At about 60 degrees north and south are found the subpolar low-pressure belts. In the Northern Hemisphere there are two such semipermanent

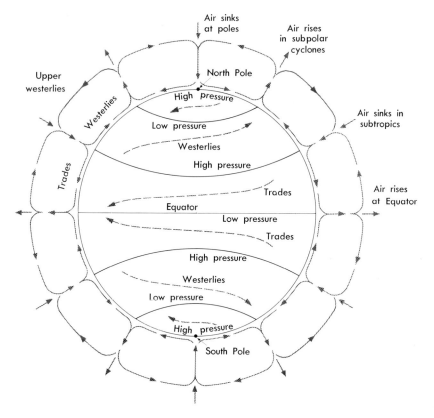

Fig. 55. The vertical and horizontal general circulation of the atmosphere.

lows: the Icelandic low in the North Atlantic and the Aleutian low in the North Pacific. North and south of these low-pressure belts are the zones of polar easterlies.

With increasing altitude above the earth, the westerly winds increase in strength and spread out over a larger range of latitudes. The shallow trade winds blowing from a predominantly easterly direction diminish in speed with increasing height; above the trade winds there are winds blowing from the west. The westerlies in the upper troposphere meander around the earth in the shape of wavelike streams of air.

The general circulation of the atmosphere is driven by the energy from the sun, and by the difference of temperature from Equator to pole. As shown in Fig. 55, warm air expands upward in the equatorial region, creating a poleward-directed pressure force at upper altitudes. In both hemispheres air flows poleward from the Equator at upper levels, only to sink in the subtropical high pressure belts and return to the Equator. This thermally driven circulation between Equator and subtropics is called the Hadley cell (after an eighteenth-century English scientists). Deflected by

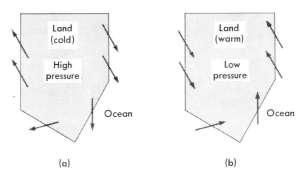

Fig. 56. The monsoon effect. (a) Winter. (b) Summer.

the Coriolis force (due to the earth's rotation), air flowing toward the Equator in the lower branch of the Hadley cell turns to the west, and thus creates the easterly trade winds. Aloft the Coriolis deflection of the poleward-moving air creates westerly winds.

A similar thermally driven circulation is found in the polar regions. There, air cooled over the poles shrinks, producing a poleward directed pressure gradient force and poleward motion aloft. The accumulation of air over the poles results in flow near the ground away from the poles and into the low-pressure belts surrounding the polar regions. Thus a circulation cell develops between the poles and the subpolar low-pressure regions. Here too the effect of the Coriolis force is to produce east winds at the surface and west winds aloft. In both cells the atmosphere acts like an engine, using temperature differences to create motion and generating enough kinetic energy to overcome the effects of friction.

While the high and low latitude cells of the general circulation can be explained as *direct* effects of heating and cooling, the middle latitude cell of westerly winds cannot. Here warm air sinks in the subtropical "highs," and cold air rises in the sub-polar "lows." This *indirect* circulation cannot generate kinetic energy. Instead the westerlies in middle latitudes are probably maintained by a continual transfer of angular momentum from the tropics effected by the large scale disturbances (cyclones and waves) in the atmosphere.

Monsoonal effects. Because of the presence of the continents, the actual wind and pressure patterns on the earth differ somewhat from the schematic presentation in Fig. 54. The influence of the continents is illustrated in Fig. 56. In winter the continents become colder than the oceans, whereas in summer the continents are warmer than the oceans. This phenomenon is due to the great heat capacity of the oceans resulting from the turbulent mixing of the ocean waters. Hence, while the land cools in winter and heats in summer, the ocean temperatures change relatively little. The air over the cold continents shrinks as it cools, and more air piles up on top. As a result high pressure develops (on the average) over the continents in winter. (An example of this trend is the great Siberian high-pressure dome.) The wind blows outward (and anticyclonically) from the continental high-pressure areas. This offshore wind is referred to as the *winter monsoon*. On the northeast coasts of the continents (e.g., at Vladivostock and New York) the winter monsoon blows from the northwest. On the west coast of India it blows from the northeast, and is cool and dry.

In summer, the vertical expansion of the heated air over the continents causes the air to flow away at high levels, and the ensuing drop in the pressure over the land leads to the development of a continental low-pressure region. Now the air blows inward (and cyclonically) from the sea into the continental low-pressure areas, giving rise to the *summer monsoon,* an onshore wind. In India the summer monsoon blows from the southwest, bringing humid air and heavy rains from the Indian Ocean.

The continents also break up the subtropical high-pressure belts into separate cells. Thus we find a subtropical high in the North Atlantic (the Bermuda-Azores high) and another in the Pacific Ocean. The circulations around these high-pressure cells drive the ocean currents in the subtropical gyres.

Smaller-scale winds. Sea breezes. The heating of the land during the day in the warm season of the year causes a local wind to blow from the sea toward the land. At higher altitudes, the air flows from the land to the sea, completing the circuit. At night the sea breeze disappears. In its place there may be a much weaker land breeze blowing from the now cooler land out to sea. The Coriolis force causes the sea breeze to rotate in a clockwise direction during the day.

The sea breeze is a direct circulation, like the Hadley cell of the tropics. Air heated over land expands vertically. A seaward directed pressure gradient force develops aloft, driving air out to sea. The air accumulates over the sea, raises the surface pressure there, and drives the sea breeze onshore near the surface; thus the circuit is complete.

This is but one of the small-scale wind systems found on earth. Others include mountain and valley winds, canyon winds, and downslope winds, all of which are associated with special features of the terrain.

Intermediate between the large-scale planetary and monsoonal circulations and the small-scale wind systems mentioned above, are the traveling cyclones and anticyclones which produce our weather.

Measurement of wind. Winds near the ground are measured with a wind vane (for direction) and an anemometer (for speed). The latter instrument may consist of revolving cups or, now more commonly, of a propeller. By wind direction is meant the direction *from which* the wind is blowing. Thus, a north wind blows from the north. The direction may also be given in degrees from north. (An east wind blows from 90 degrees, a south wind from 180 degrees, etc.) Wind speed is expressed in miles per hour or in knots.

Winds at upper levels in the atmosphere are determined by tracking helium- or hydrogen-filled balloons which are carried along with the wind. If the tracking is done visually (optically) with a telescopic device called a *theodolite,* the observations are referred to as pilot-balloon observations (*pibals*). Winds tracked electronically (by radio-direction finding or radar) are called *rawins* (radio winds). This latter method may be used to measure winds up to an altitude of about 30 kilometers. Winds at altitudes above the ceilings of the balloons can be measured by means of rockets that eject smoke puffs or radar-reflecting targets at high altitudes.*

4. Weather. By *weather* we generally mean the products of condensation in the atmosphere (fog, clouds, rain, snow, hail, sleet, drizzle) as well as some of nature's more violent manifestations, such as thunderstorms, lightning, and destructive winds. At present, we confine our attention to the physical processes in the atmosphere that produce clouds and precipitation. We shall try to understand these phenomena in relation to weather maps, and we shall also attempt to gain some insight into the practice of weather forecasting.

Condensation and precipitation. Weather begins with the water vapor in the air. By processes which are discussed below, water vapor may reach a state of saturation, or even supersaturation, whereupon it condenses on condensation nuclei in the air, and tiny invisible ice crystals or droplets form. These droplets and crystals grow into cloud drops and crystals. Supported by vertical air currents, the clouds hang in the atmosphere, and are carried along by the wind. In many clouds the water soon evaporates, returning to the vapor phase as the cloud disappears. But in some clouds the drops and crystals grow into raindrops or snowflakes, and fall to the earth as precipitation.

Saturation (and supersaturation, which means a relative humidity greater than 100 percent) may result either from addition of water to the air by evaporation or by cooling of the air. The evaporation process is important only in the production of certain types of fog, e.g., the fog that develops during the night after a day of rain, or the "steam fog" that forms when very cold air passes over warm water bodies. Evaporation cannot lead directly to the precipitation of rain or snow.

There are several ways in which the vapor in the air may be brought to a state of saturation (temperature equal to the dew point) by cooling. At night, in the absence of insolation, the air near the ground loses heat to

---

* See, in particular, reference 7.

space by infrared radiation. This heat loss can cause ground fog, but not rain. When warm air passes over cold land or water, the air cools by heat exchange, and the drop in temperature may again produce only fog, but not rain. Finally, the air may be cooled by expansion. This is the most important weather process, and the only one that leads to the development of clouds and precipitation.

When a bubble of air rises, it enters a region of lower atmospheric pressure. As in a balloon, the air in the bubble expands, and in expanding, does "work" by pushing outward. Work requires the expenditure of energy, and this energy is provided by the *internal energy* (associated with the molecular motion) of the air. The expenditure or loss of internal energy causes the air to cool, for the internal energy is proportional to the air's (absolute) temperature. Thus the rising air cools by expansion. Note that this cooling occurs without any transfer of heat out of the air, for the air is a very poor conductor of heat (and a good insulator). A process in which the temperature changes without heat transfer is called an *adiabatic process*. Rising dry air cools (by expansion) adiabatically at the rate of about $10°C$ for every kilometer ($6°F$ for every 1000 feet) it ascends. Sinking air warms (by compression) in the adiabatic process.

Weather is caused by vertical motion of air, for only the adiabatic cooling of rising air can lead to significant cloud formation and precipitation. When air rises, its temperature falls faster than its dew point, and at some level, called the *condensation level,* the temperature overtakes the dew point, and condensation occurs. The cloud base forms at the condensation level, and the cloud grows by condensation above that height. The cloud particles remain suspended in the air, supported by the vertical currents that produced them.

For the cloud water (drops or crystals) to fall to the ground as precipitation, the cloud particles must grow. Large drops fall faster than small drops in a viscous medium such as air because of the effect of *drag* (friction). If the drops grow large enough, they will fall faster than the air is rising, and thus fall to the earth as precipitation. The drops grow principally by two processes: coalescence of drops or an ice-crystal process. Coalescence takes place when larger drops capture smaller drops falling past them in the clouds. Th ice-crystal process results from the fact that water vapor condenses more readily on ice than on liquid water when both are present in a cloud whose temperature is less than $0°C$. Liquid water may exist in a cloud at temperatures well below "freezing," even at the very low temperature of $-40°C$. Such water is said to be *supercooled*. Icing of aircraft in clouds is caused by the freezing of this supercooled water. The ice crystals grow at the expense of the liquid water in the supercooled cloud.

Supercooled clouds can be converted into ice crystals by *cloud seeding*. This procedure consists of dropping pellets of very cold material (e.g., dry ice, which is solid CO_2) into the cloud, or of injecting freezing nuclei, such as microscopic crystals of silver iodide, into it. The latter are released as smokes. Freezing nuclei or dry ice cause the supercooled water to freeze, and initiate the growth of ice crystals. The seeded clouds thus precipitate snow, so that holes or furrows may be cut in the cloud by seeding. It has been claimed, and with some evidence, that under certain favorable conditions the amount of precipitation may be increased artificially by cloud seeding.

Causes of vertical motion. Since almost all weather is caused by vertical (upward) motion, we can seek to understand weather phenomena by examining the causes of these vertical motions.

Weather phenomena occur on different scales. A cumulus cloud growing into a thunderstorm cloud (cumulonimbus) is an example of a small-scale weather phenomenon. A broad sheet of altostratus cloud, covering perhaps a half-million square miles, with rain falling over many states, is an example of a large-scale weather process. In general, the upward motions are slower in the large-scale weather systems than in the small-scale systems. As a result, the small-scale cloud systems tend to develop vertically (cumulus clouds), whereas the large-scale systems develop horizontally, in sheets (stratus clouds). The vertical motions in cumulonimbus clouds (thunderheads) may occur with speeds exceeding 50 miles per hour, and may be strong enough to produce violent turbulence for aircraft. These strong updrafts also often lead to the formation of hail. Because strong upward currents are required to support a hailstone while it is growing, hail is observed only in connection with thunderstorms.

It should be noted that the sinking motion of air (subsidence) causes the temperature of the air to rise above the dew point, resulting in the evaporation of clouds and clearing weather.

Upward vertical motions are produced by the effects of topography, by convergence of horizontal air currents, by the upgliding of one air mass over another, and by convection. Each of these factors leads to a different kind of cloud and weather phenomenon.

The effect of topography (e.g., a mountain) on weather is illustrated in Fig. 57. The air blowing toward a mountain (e.g., a west wind blowing against the Sierra Nevada Mountains) is forced to rise on the windward side, but sinks on the leeward side. As a result of adiabatic cooling, *orographic* (mountain) clouds and precipitation develop on the windward side of the mountain (where dense forests grow), whereas adiabatic warming on the leeward side produces a dry "rain shadow" (desert).

We have already seen how air currents converge into a cyclone (Fig. 52b). This piling-up of the air must be accompanied by a rising air stream which flows out of the cyclone at high altitudes, as shown in Fig. 58. Otherwise, the pressure would rise in the low-pressure center. The upward vertical motion accompanying the convergence of air cur-

Fig. 57. Vertical motions and orographic weather produced by a mountain barrier (vertical cross section).

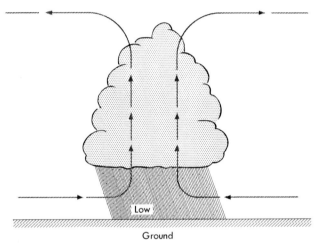

Fig. 58. A vertical cross section through a cyclone, showing convergence of horizontal air currents and resultant vertical motion and weather.

rents causes adiabatic cooling, condensation, clouds, and precipitation in cyclones and in other locations where the air streams converge in the lower levels of the atmosphere. Of course, the air must be humid as well as rising for the quantity of precipitation to be appreciable.

Upgliding takes place when a mass of warm air encounters a wedge-shaped mass of cold air, as shown in Fig. 59. The warm air slides up over the cold mass, producing widespread cloud layers and precipitation. The boundary between the two air masses is called a *front*.

Convection is the vigorous vertical motion on a small scale which is responsible for cumulus clouds, showers, and thunderstorms. In extreme cases, convection may produce tornadoes. These are small, funnel-shaped vortices which contain the most violent winds on earth. Convection occurs when the air is in an *unstable* state, i.e., in a precarious state of balance, so

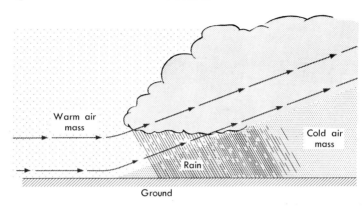

Fig. 59. Upgliding of warm air over a wedge of cold air. The boundary between the air masses is a *front* (vertical cross section).

that even the smallest impulse will initiate rapidly accelerating vertical motions. The air is unstable when the lapse rate is large, i.e., when the decrease of temperature with height exceeds a certain critical value. The air may become unstable through heating at the bottom or through cooling in the upper layers. On hot summer days, due to the heating at the bottom, bubbles of heated air rise from the ground, accelerate upward, and produce shower clouds by adiabatic cooling and condensation. This upward movement, which may lead to the growth of cumulus clouds and the formation of afternoon showers and thunderstorms, is sometimes called thermal convection. But convection may also occur at night, when the ground is not being heated. In many instances, convection, which may produce violent thunderstorms, is caused by cold air flowing in at upper levels over the warmer air below. This phenomenon leads to instability and overturning of the air. The lifting of air caused by the passage of a front may also set off a convection process, either by day or night.

Cyclones, fronts, and weather forecasting. Aside from the visual and instrumental observations of the local weather, the principal tool of the weather forecaster is the weather map on which are plotted the simultaneous observations of hundreds of weather stations. Weather maps, also called *synoptic maps,* give the forecaster a bird's-eye view of the weather over a large area at a given moment.

Except for scattered reports from ships, most of the weather data received by weather forecasters comes from land areas. From vast areas of the earth, no weather information at all is received. With meteorological satellites, such as the Tiros satellite, it is now possible to fill these gaps in the large-scale weather picture. Television photographs of clouds seen from above the atmosphere, and measurements of infrared radiation emitted from the earth, are obtained by the satellites and transmitted to meteorological centers by radio, helping to complete the synoptic weather map.

The forecaster integrates and summarizes the myriad of weather observations by analysis, i.e., he identifies the different kinds of air masses on the map, locates the fronts, or boundaries, between the air masses, delineates areas of precipitation, and indicates the distribution of pressure by drawing isobars, which are lines of equal pressure. The analyst also notes the cyclones (lows) and anticyclones (highs) on the map.

The principal air masses are cold, dry continental polar air (e.g., from Canada), cold, humid maritime polar air (e.g., from the North Pacific Ocean), and warm, humid maritime tropical air (e.g., from the Gulf of Mexico). These air masses interact along fronts to produce much of our weather.

The principal fronts are cold fronts and warm fronts. Along cold fronts, the cold air is advancing as the warm air retreats, and as a result, there is a narrow band of showers and thunderstorms along the front. Along a warm front, the warm air is advancing while the cold air retreats, and this situation produces a broad area of clouds and precipitation ahead of the front.

As shown in Fig. 60, the fronts are closely related to the cyclones. Figure 60 shows a typical middle-latitude cyclone containing cold and warm fronts. Cyclones of this kind are called *wave cyclones* because they form as wavelike wrinkles on a front. As the wave grows in amplitude, the pressure at the peak of the wave falls, and the winds intensify. Part of the front moves as a cold front, while another part moves as a warm front, due to the wind pattern. At the same time, the wave travels along the front, usually from west to east, carried by deep currents of air, aloft.

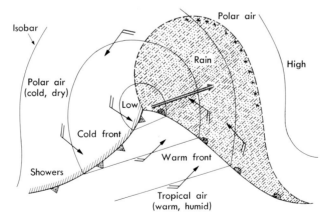

Fig. 60. Typical middle-latitude cyclone showing isobars (thin lines), fronts, precipitation, wind arrows, and the motion of the cyclone (double arrow).

CHAP. 4]   SOLAR RADIATION AND CLIMATE   123

Hurricanes and typhoons are very intense tropical cyclones. Although they are initially small, they may grow to very large size as they mature. The wind speeds in these storms exceed 33 meters per second (65 knots) in places, but the cyclones themselves move slowly. They are called hurricanes in the Atlantic Ocean, typhoons in the Pacific. The relatively calm central region of the storm, about 20 to 40 kilometers in diameter, is called the *eye*. Tropical cyclones, including hurricanes and typhoons, do not contain fronts, and are more symmetrical than mid-latitude storms. They also contain but a single air mass, tropical maritime air, and produce more violent winds and heavier rain.

The basic principle of weather forecasting from the synoptic weather map is *weather moves*. By studying the past movements of the cyclones, fronts, and highs, the forecaster is often able to predict what the weather map will look like tomorrow. These *prognostic* weather maps are constructed by moving the weather systems according to their past behavior, by calculating the movements of systems from theory, by using statistics based on past records, and by dynamical computations carried out on high-speed, electronic computing machines. Maps of the air currents in the upper atmosphere also play an important role in forecasting, for cyclones are carried along by these currents like vortices in a stream. After the prognostic weather map has been drawn, the forecaster must interpret its significance in terms of the local weather. Forecasting is not a complete or exact science, and errors are to be expected because of the incomplete state of our knowledge of weather phenomena and, especially, because of the complexity of the atmosphere.*

5. Solar radiation and climate. Climate is the average state of the atmosphere. The climates of the earth and the variation of climate with latitude can be explained, to a first approximation, by the global distribution of insolation. Similarly, we can account for the seasons by considering the motion of the earth about the sun, the tilt of the earth's axis, and the resulting seasonal variation of insolation. For a complete survey of the world's climates, we must also consider the distribution of continents and oceans, the warm and cold ocean currents, the influence of the mountains, and the development and movement of cyclonic storms.

The sun and its radiation. The sun is a glowing ball of hot gas whose energy is derived from a continuous thermonuclear reaction, a process of fusion in which hydrogen is transformed into helium and mass is converted into energy. Every second, the sun sends out an amount of energy equivalent to that which would be produced by burning 25 billion billion pounds of coal. The vertical structure of the sun is shown in Fig. 61. Here we see that the greatest part of the sun, its interior, where the

---

* See, in particular, reference 7.

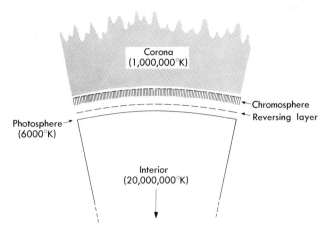

Fig. 61. Structure of the sun.

thermonuclear reaction takes place, is probably as hot as 20 million degrees celsius.

The surface of the sun is known as the *photosphere* (Gr. *phos,* light). The light reaching us from the sun emanates from the photosphere, so that this is the part of the sun we see. The temperature of the photosphere is about 6000 degrees kelvin (absolute, or kelvin, temperature is equal to degrees celsius + 273.) The radiation emitted by the photosphere corresponds almost, but not exactly, to the theoretical blackbody radiation of an ideal or perfect radiator. The radiation from such a body can be calculated from the physical laws of radiation. One of these, the Stefan-Boltzmann law, states that the rate of emission of energy by a blackbody increases in proportion to the fourth power of the absolute temperature of the body (T^4). A method for determining the sun's temperature is to measure the energy emitted by the sun (as received at the earth), and to calculate what the temperature of the radiating blackbody must be.

The sun emits a continuous spectrum of electromagnetic radiation, i.e. all wavelengths of light, including invisible infrared and ultraviolet radiations. However, there is more energy in some wavelengths than in others. The law regarding the intensity of the blackbody radiation in any wavelength is known as Planck's law. A special law which indicates in which wavelength the *maximum* radiant energy is contained is known as Wien's law. According to this law, the wavelength of maximum intensity varies *inversely* as the absolute temperature. Thus hot bodies radiate predominantly short-wave radiation while cold bodies radiate mainly long-wave radiation. The sun emits most of its energy in the short-wave region of the solar spectrum, i.e., as visible light. We can determine the temperature of the sun by measuring the solar radiation received at the earth with a spectrometer, noting the wavelength (color) in which the maximum energy is contained, and applying Wien's law. This temperature, which is known as the color temperature of the sun, is a little less than 6000°K (kelvin).

Dark regions are occasionally seen in the photosphere. These are the *sunspots,* cooler regions of the photosphere from which the radiation is less than normal. However, sunspots are often accompanied by bright regions of the sun (faculae), which may more than compensate for the cool sunspots.

Above the photosphere lies the solar atmosphere, consisting of three regions: the reversing layer, the chromosphere, and the corona. The first is a layer of cool gas that absorbs certain wavelengths of sunlight (the Fraunhofer lines). The chromosphere, composed mainly of hot hydrogen, emits a red glow that can be seen only during an eclipse of the sun. The corona is a tenuous envelope of extremely hot gas, also seen only during an eclipse, extending millions of miles from the sun. Radiations from the chromosphere and corona probably have an influence on the earth's upper atmosphere, but do not appear to affect the climate of the lower atmosphere.

Insolation. As far as we know the sun emits energy at a nearly constant rate. The intensity of this radiant energy diminishes with increasing distance from the sun, in accordance with the inverse-square law. When the earth is at its average distance (150 million kilometers) from the sun, a surface held perpendicular to the sun's rays at the top of the earth's atmosphere receives almost exactly two calories per square centimeter per minute. (A calorie is the amount of heat needed to raise the temperature of one gram of water by one degree celsius.) This quantity of energy is known as the *solar constant.*

The intensity of the solar radiation falling on any portion of the earth's surface, however, is not contant. As shown in Fig. 62, there are two reasons why the radiation falling on a horizontal surface varies with the

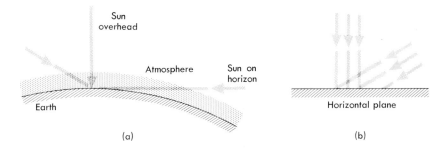

Fig. 62. (a) Variation in the intensity of insolation falling on a horizontal surface resulting from variations in the elevation of the sun. (b) Effect of solar elevation on areas irradiated by solar beam.

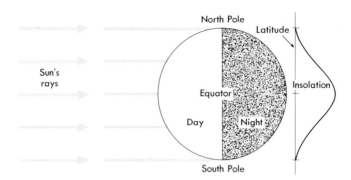

Fig. 63. Radiation from the sun falling on the horizontal surface of the earth (average annual conditions). Note variation in elevation of the sun with latitude.

elevation of the sun. First, when the sun is low on the horizon, its radiation passes through a thicker layer of atmosphere than when the sun is overhead, and the radiation is therefore more attenuated by absorption and scattering. Secondly, when the sun is low in the sky, its radiant energy is spread over a larger area of the earth's surface than when it is high, so that the radiation *per unit area* is less.

The average insolation per unit area falling on the earth's surface each year is greatest at the Equator and smallest at the poles. The reason for this variation is illustrated in Fig. 63, which shows that the average solar elevation decreases from a maximum at the Equator to a minimum at the poles. As a result of this distribution of insolation, the average annual temperature of the earth's surface and of the air near the ground decreases from the Equator to the poles. (The earth is warmed by the insolation which it absorbs. The earth, in turn, heats the air by long-wave infrared radiation, which the earth emits and the air absorbs.)

The seasons. The seasons result from the tilt of the earth's axis relative to the plane of the ecliptic (the obliquity of the ecliptic). Figure 11 illustrates the effect of the constant tilt of the earth's axis in space on the length of the day.* At the equinoxes, the day is exactly 12 hours long at all latitudes in both hemispheres. At the winter solstice (about December 21), which marks the beginning of the winter season in the Northern Hemisphere, the North Pole tilts away from the sun, causing the length of the day to be at a minimum everywhere in the Northern Hemisphere and at a maximum everywhere in the Southern Hemisphere on that date. Six months later, at the summer solstice, when the North Pole tilts toward the sun, the length of the day is at a maximum at all

---

* In the present context, the term "length of day" refers to the time interval from sunrise to sunset.

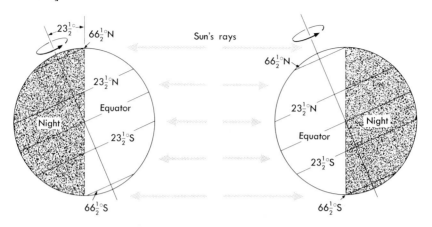

Fig. 64. The tilt of the earth's axis at the summer and winter solstices and the effect on the inclination of the sun's rays to the horizontal plane at different latitudes. (a) Winter solstice (December 21). (b) Summer solstice (June 21).

latitudes in the Northern Hemisphere. The length of the day increases from 12 hours at the Equator (always) to 24 hours at and north of the Arctic Circle (latitude $66\frac{1}{2}$ degrees north) at the summer solstice, but decreases from 12 hours at the Equator to zero at and south of the Antarctic Circle (latitude $66\frac{1}{2}$ degrees south). At the winter solstice, the situation is reversed in the two hemispheres. The long summer days and short winter days represent one factor in the seasonal variation of air temperature on the earth.

The second solar factor controlling our seasons is the variation in the elevation of the sun, as illustrated in Fig. 64. Here we see that at the summer solstice the $23\frac{1}{2}$-degree tilt of the earth's axis causes the sun to appear directly overhead at noon at latitude $23\frac{1}{2}$ degrees north, the Tropic of Cancer, with the result that this latitude receives the maximum insolation per unit area. At the winter solstice, the maximum insolation per unit area is received at latitude $23\frac{1}{2}$ degrees south, the Tropic of Capricorn. Due to the migration of the sun back and forth across the Equator, the belt of maximum temperature, the so-called thermal equator, also migrates following the sun. (We may note that Fig. 63 represents conditions at the equinoxes.) It is obvious from Fig. 64 that the insolation received by the Northern Hemisphere is greatest at the summer solstice, while the insolation received by the Southern Hemisphere is greatest at the winter solstice.

The annual variation in temperature from winter to summer is a consequence of the annual range of insolation. Since the fluctuations in insolation are at a minimum at the Equator, this region is characterized by an almost total absence of annual temperature variations and seasons (except wet and dry). In general, at the Equator the temperature of the "warmest" month is rarely more than 1°C warmer than the temperature of the "coldest" month.

The annual range of insolation increases with latitude, and consequently the annual temperature range, the difference in temperature between winter and summer, increases as we move poleward from the Equator.

It should be noted that the annual course of air temperature lags behind the annual course of insolation by about a month, so that the coldest and warmest periods occur later than the winter and summer solstices.

Effects of continents and oceans on climate. If the earth were a uniform body (e.g., completely covered with water), its surface temperature would be controlled entirely by the distribution of insolation. Since, in the absence of clouds, the daily insolation varies only with season and latitude, the average temperature pattern would be a simple one. Isotherms (lines of equal temperature) would be parallel to the latitude circles, with temperature decreasing from the Equator poleward (on the average for the year). In the Northern-Hemisphere summer, the thermal equator would be found north of the Equator, whereas in the Northern-Hemisphere winter, the thermal equator would be south of the Equator. The decrease of temperature from Equator to pole would be at a maximum in the winter hemisphere, and at a minimum in the summer hemisphere. (In the summer hemisphere there would, in fact, be very little variation of temperature with latitude.) The coldest place on earth would be the pole of the winter hemisphere.

That this planetary climate does not exist in the simple form indicated above is due in large part to the differences between the continents and oceans. The energy absorbed by land is used to heat a thin layer of the earth, whereas, due to the overturning of the ocean waters, the energy absorbed by the ocean heats a thick layer. The temperature change is proportional to the heat absorbed *per unit mass*. As a result, the land heats up rapidly in summer, while the oceans do not. Conversely, when the oceans lose energy in winter, this heat comes from a large mass, whereas the land loses its heat from a small mass. Consequently the land cools rapidly in winter, and the oceans do not. Furthermore, a significant amount of the energy absorbed by the oceans in summer is used for evaporation, and therefore does not heat the water. This is another factor contributing to the near constancy of the ocean's temperature. Thus the continents are cold regions in the winter hemisphere and hot regions in the summer hemisphere relative to the oceans, and these characteristics become more pronounced as one moves toward higher latitudes. Figure 65 illustrates schematically the effect of continents on the worldwide temperature distribution.

CHAP. 4] SOLAR RADIATION AND CLIMATE 129

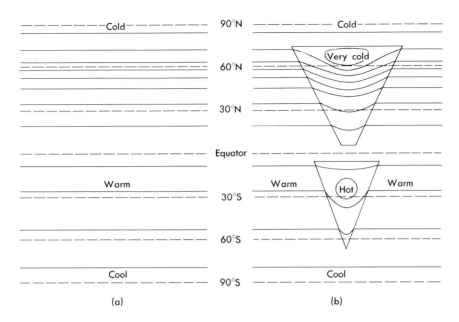

Fig. 65. (a) "Planetary" temperature distribution at sea level as it would appear on a uniform earth at the winter solstice. (b) Effects of continents on the temperature pattern. Solid lines are isotherms.

The continentality effect described above leads to the following conditions:

(1) The largest annual temperature range between winter and summer is found over the interiors of the large continents in middle or high latitudes. Continental climates are said to exhibit severe climatic "stress," with very cold winters and very hot summers.

(2) The lowest winter temperatures in the Northern Hemisphere are found not over the North Pole but over Northeastern Siberia. In the Southern Hemisphere, however, where a major continental mass surrounds the South Pole and most of the hemisphere is ocean, the lowest temperature is found near the South Pole.

(3) The contrast between continents and oceans produces very large gradients of temperature from east to west, as well as from north to south. The temperature along any latitude decreases from the ocean to the continents in winter, and increases from the ocean to the continents in summer.

(4) East coasts of continents in the belt of westerly winds have continental climates, whereas west coasts in these latitudes have maritime climates. (Compare New York's annual temperature range of 43°F from January to July with San Francisco's annual range of less than 11°F.)

We have not discussed a number of important climatic influences. One of these is mountains. The temperature decreases with increasing altitude. Thus a mountain-top city on the Equator may have an average annual temperature lower than a city at sea level in middle latitudes.

World rainfall. The worldwide distribution of rainfall exhibits an extremely complicated pattern. For example, precipitation is influenced by mountains: heavy rain falls on the windward side of a mountain, light rain on the leeward side. Rainfall is also affected by the tracks of moving cyclones and many other factors. We present here only a generalized picture of the distribution of rainfall with latitude.

Factors which favor heavy rainfall are high humidity, convection (unstable air), convergence of air streams, cyclones, fronts; those which lead to low rainfall are low humidity (and low temperature), subsidence, diverging winds, and anticyclones.

At the Equator the air is warm and humid. Equatorial air is the most heavily moisture-laden air in the world. The trade winds converge at the doldrums or the equatorial low, also known as the intertropical convergence zone. The equatorial air is unstable, and lifting of this air releases violent convection, heavy showers, and thunderstorms. As a result, the Equator is the latitude with the heaviest rainfall in the world.

Rainfall decreases poleward from the Equator, reaching a minimum in the subtropical high-pressure belt of both hemispheres. Here the wind diverges from the anticyclones, the air sinks and warms, and clouds evaporate. The air is dry and stable. The result is that rainfall minima occur at about 25 degrees north and 25 degrees south.

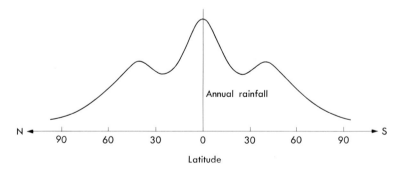

Fig. 66. Variation of annual rainfall with latitude.

As we proceed poleward in the westerlies, rainfall increases. This is due partly to the increase in the number of cyclones passing through the westerlies. Also, this is the region of the *polar front,* the corrugated boundary between the polar and tropical air masses, along which the air masses converge and rise, producing clouds and precipitation. Cyclones form on the polar front, and travel along it like waves. The result is that secondary rainfall maxima are found in the zone of westerlies at latitudes of 45 degrees north and 45 degrees south, approximately.

Poleward of the mid-latitude rainfall maxima the precipitation decreases steadily, reaching minima at the poles. This is largely due to the fact that the total amount of water vapor in the cold polar air is so low that even if other favorable conditions were present, no large amounts of precipitation could occur in this air.

The world rainfall distribution is illustrated schematically in Fig. 66.*

6. The upper atmosphere. The exploration of the upper atmosphere of the earth by rockets and satellites is advancing so rapidly that it is impossible to write an up-to-date account of this branch of atmospheric science. We shall therefore be content to present a background review of some of the basic facts that are known about the structure and physical processes of the upper atmosphere, indicating, if possible, where rockets or satellites have added to our knowledge in this field.

In this section we discuss the phenomena of the ionosphere, including auroras, geomagnetic disturbances, and other effects of anomalous emissions from the sun. We also mention briefly meteors, cosmic rays, and radiation belts.

Above the stratosphere. An important physical change in the atmosphere begins about 60 kilometers above the earth, in the upper mesosphere. In this region, known as the ionosphere, the concentration of free ions and electrons increases to such high levels that it is possible to think of the air as an electrical conductor. Although the ionosphere begins below the mesopause, maximum ionization is found at an altitude of more than 100 kilometers, in the thermosphere. The ionization of the air in the ionosphere is produced by x-ray and ultraviolet radiation from the sun.

The existence of the ionosphere was first inferred by Balfour Stewart in 1883 from observed variations in geomagnetism. Stewart guessed correctly that the fluctuations in the earth's magnetic field seen on magnetograms were caused by the movements of a conducting layer (the ionosphere) across the earth's permanent magnetic field. These movements initiate electrical currents in the upper atmosphere, and induce magnetic disturbances in the earth.

After Marconi's success in 1901 in sending radio signals across the Atlantic Ocean, Kennelly (in the United States) and Heaviside (in England) independently suggested that it was the existence of an electrical conducting layer (the ionosphere) which made this feat possible by acting as a reflecting surface for radio waves, returning them to the earth over long distances (Fig. 67).

---

* See, in particular, references 7 and 11.

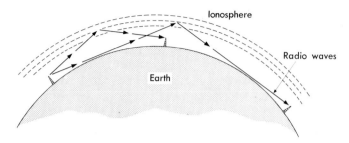

Fig. 67. Long-distance radio communication made possible by the existence of the ionosphere.

It is now recognized that there are at least three principal ionized *regions* in the ionosphere. These are the D-region, below 90 kilometers, the E-region, which extends from 90 to 160 kilometers, and the F-region, which includes all ionization above 160 kilometers. Within these regions are the *reflecting layers:* the D-layer at 80 kilometers, the E-layer at 120 kilometers, the $F1$-layer at 200 kilometers, and the $F2$-layer at 300 kilometers above the earth.

Above the $F2$-layer, where the temperature is probably more than 750°C, there is an abundance of atomic oxygen. At still higher levels, where the air temperature may be as high as 1700°C, the light gases helium and hydrogen are probably important constituents of the air.

The atmosphere itself must extend well above the $F2$-layer, because on some rare occasions the aurora borealis (northern lights) has been observed to altitudes of 1000 kilometers. It now appears rather certain that above the layer of atomic oxygen, which extends to about 1000 kilometers, there is a layer of helium. The helium layer extends from about 1000 to 1500 kilometers. Above 1500 kilometers the dominant gas is the lightest element, hydrogen. At about 10,000 kilometers above the earth the density of the hydrogen layer fades into the density of interplanetary space, and the earth's atmosphere may be said to "end."

The aurora is produced in the upper atmosphere of the earth by particles (electrons and protons) which have been ejected by the sun. Upon entering the upper atmosphere, these electrical particles excite the gases there, causing them to radiate light. (A similar process takes place in a neon or fluorescent light tube.) The magnificent, shimmering, colored lights seen in the arctic sky thus originate in the atmosphere and tell a good deal about the composition of the ionosphere. Measurements of aurorae have shown that the phenomenon occurs most frequently at an altitude of about 100 kilometers, in the E-region.

High above the F-region, in what must still be considered a part of the earth's atmosphere, rockets and satellites have brought about the discovery of the so-called Van Allen radiation belts (Fig. 68). These belts

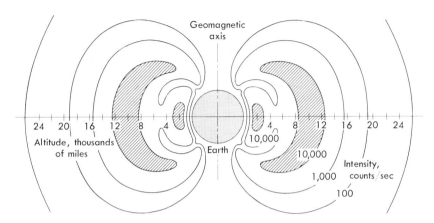

Fig. 68. The Van Allen radiation belts.

are regions occupied by high-energy particles (electrons and protons) which are capable of producing intense ionization, including radiation burns on space travelers. The origin of these belts is still uncertain, but it is likely that they have two sources: cosmic rays and streams of gas from the sun. Cosmic rays are believed to produce the inner (lower) belt, the bottom of which is found about 3000 kilometers (2000 miles) above the earth, while streams of ionized gas from the sun may account for the outer belt. The latter appears to extend from about 16,000 kilometers out to about 100,000 kilometers above the earth, with its "center of gravity" at about 25,000 kilometers. The intensity of the radiation in the Van Allen belts is greatest over the geomagnetic equator, and diminishes toward the poles, whereas auroras occur most frequently in concentric rings around the geomagnetic poles and diminish in frequency toward the Equator.

The Van Allen belts and the associated regions of the very high atmosphere are now referred to collectively as the *magnetosphere* because of the fact that the physics of this region is dominated by the earth's magnetic field. Electrically charged particles from the sun, in the form of "plasma clouds" containing both negative and positive charges, are hurled into space by eruptions in the solar atmosphere. The particles reaching the outer Van Allen belt are trapped in the earth's magnetic field. They spiral back and forth across the geomagnetic equator along the lines of magnetic force, from one hemisphere to the other. An intense cloud of plasma may cause a cascade of the trapped particles toward the "horns" of the Van Allen belts at about 65° north and south magnetic latitude. Here, where the Van Allen belt discharges its burden of charged particles, the auroras occur. The Van Allen belt refills again quickly after its radiation has leaked off to produce the auroras, so that the first auroral display may be followed shortly by others.

The exact mechanism by which the solar plasma cloud emitted by a solar flare induces the outer Van Allen belt to empty its radiation into the auroral zone is not completely known. However, it is known that the plasma cloud carries its own magnetic field with it. This field is strong enough to deflect cosmic rays away from the earth, so that during a magnetic storm the cosmic-ray intensity on the earth decreases. It is probable that the magnetic field of the plasma cloud disturbs that of the magnetosphere in such a way as to permit the trapped radiation to leak off into the ionosphere in high latitudes, where the auroras occur.

How we observe the upper atmosphere. Information about the state of the upper atmosphere has been obtained by direct and indirect measurements. Before the development of rocket vehicles (and satellites), all information was obtained indirectly from observations of meteors and anomalous sound propagation, and from theoretical calculations.

Meteors (or shooting stars) are tiny pebbles of stone (with some metal) whose size rarely exceeds that of grains of sand. They occur in swarms, which move in regular elliptical orbits about the sun. When a meteor swarm intersects the earth's orbit, the meteors enter our atmosphere, and a meteor shower is observed. Friction between the air and the fast-moving meteor generates enough heat to bring the meteor to incandescence for a few seconds. Most of the meteors we see as shooting stars evaporate in the upper atmosphere. The few large meteors which do not evaporate completely reach the earth as meteorites. The appearance and disappearance of meteors provide information about the density of the atmosphere. In 1923, Lindemann and Dobson showed from a study of the altitudes of visible meteors that there must be a layer of relatively high density (cold air) at a height of about 80 kilometers (now known as the mesopause) and a relatively low-density (warm) layer at about 50 kilometers (the stratopause).

About the same time, F. J. W. Whipple proved the existence of the warm stratopause layer by studying the skipping of sound waves over great distances. This phenomenon of anomalous audibility had been known since 1901, when the sounds of Queen Victoria's funeral guns were heard hundreds of miles away. Whipple showed that this "leapfrogging" of the sound was caused by the refraction of the sound waves in a warm layer of air 50 kilometers above the earth. The sound rays were bent back to earth in a manner similar to the reflection of radio waves by the ionosphere.

The existence of the cold mesopause, 80 kilometers above the earth, was also guessed at by meteorologists who observed clouds at that altitude. These high clouds, seen only at night by the reflected light from the sun, are called *noctilucent* clouds. (It is now generally believed that noctilucent clouds are composed of dust rather than water or ice, and hence may not really tell us anything about the air temperature. This, however, is still a controversial question.)

The vertical structure of the atmosphere was also deduced, in 1937, from the theory of atmospheric tides. The tide in the atmosphere, which is seen as a semidiurnal (12-hourly) oscillation in the air pressure on a barograph, can be explained by the existence of a cold mesopause.

Direct measurements of the atmosphere above the balloon ceiling of about 30 kilometers are made with rockets. Among the many ingenious techniques for measuring pressure and temperature with rockets, we will mention only the rocket-grenade method, in which grenades are exploded at different altitudes during the rocket's flight. The speed of the sound from the explosion is measured at the ground, and the air temperature is calculated from the sound speed, which depends on the temperature and chemical composition of the air. Measurements of ionization, meteor impacts, solar radiation, and cosmic rays are but a few of the uses to which scientific rockets and satellites are being put today.

Cosmic rays. The mysterious high-energy radiations from space known as cosmic rays are responsible for the carbon-14 in the upper atmosphere, and probably for the inner Van Allen radiation belt. They also help to ionize the air near the ground.

Cosmic rays are protons, helium nuclei (alpha particles), or heavy atomic nuclei whose extraterrestrial origin was first discovered in 1911 by Hess. In colliding with atmospheric atoms and molecules in the upper air, the primary cosmic rays produce showers of secondary cosmic rays which penetrate deeper into the atmosphere. Neutrons released by cosmic-ray bombardments collide with nitrogen atoms and transform the nuclei of the nitrogen atoms, converting some of the nitrogen into the radioactive isotope of carbon (carbon-14). This isotope in turn combines with oxygen to produce radioactive carbon dioxide in the upper atmosphere, and it is this carbon dioxide which enters the biosphere and can then be used for age determination of dead matter by radiocarbon dating.

The ionization burns caused by cosmic rays will probably constitute a hazard to space travelers, unless adequate shielding is provided.

Storms on the sun and their effects on the earth's atmosphere.* Long-distance radio communication depends on the reflection of radio signals by the ionosphere. The ionosphere not only reflects radio waves (by refraction), but also absorbs them. The E- and F-layers reflect radio waves back to the earth. The D-layer is ineffective as a reflector for most radio waves, but it does absorb the energy of the radio waves. Increased D-layer ionization therefore weakens radio signals and interferes with radio communication.

---

* See Section 5.

During the day, the ultraviolet light from the sun intensifies the D-layer, but at night the D-layer is absent. Therefore, short-wave radio communication over long distances is better at night than during the day.

Besides the diurnal and seasonal variations in ionization and radio-signal strength, there are also long-period variations associated with the sunspot cycle. The number of sunspots reaches a peak about every 11 years, i.e. the sunspot cycle has an average period of 11 years. It appears that an increase in the number of sunspots enhances the ionizing ultraviolet radiation, with a corresponding effect on radio communications.

The most spectacular and least predictable of all the variations in the ionosphere are those associated with erratic disturbances on the sun. The most violent of these disturbances are *solar flares,* intensely heated regions of the sun which hurl enormous quantities of high-energy particles explosively out of the chromosphere. In addition to these particle streams, called corpuscular radiation, the flares also produce enhanced ultraviolet radiation.

The wave (ultraviolet) radiation from the flare travels at the speed of light, reaching the earth the moment the flare is seen (eight minutes after the event has occurred on the sun). The particles travel more slowly, and take a little over a day (26 hours) to reach our atmosphere. The flare thus produces one immediate and one delayed effect, since both wave and particle radiations ionize the upper air.

At geomagnetic observatories, the arrival of these anomalous radiations in the ionosphere is indicated by fluctuations in the earth's magnetic field.* Intense worldwide magnetic disturbances, known as *magnetic storms,* follow the corpuscular emissions from the sun, often lasting for several days. Radio fadeout, due to increased D-layer ionization, and even radio blackout (especially in the Arctic), due to complete failure of the F-layer to reflect signals, are two of the consequences of ionospheric disturbances. Solar flares may be followed by brilliant auroras. These too are caused by the particles which are ejected from the storm on the sun, enter the upper atmosphere, excite it, and cause it to glow.

The study of the eruptions on the sun and their consequences in our atmosphere is being carried out by astronomers, physicists, and meteorologists at solar astronomical observatories, at radio communication laboratories, and at geomagnetic observatories, and with rockets and satellites. This field is one of the frontiers of science.†

---

* See Chapter II, Section 5.
† See, in particular, references 9, 11, and 12.

References

1. D. R. BATES, "Composition and Structure of the Atmosphere," in *The Earth and Its Atmosphere*. New York: Basic Books, 1957, pp. 97–112.
2. E. T. EADY, "Climate," in *The Earth and Its Atmosphere,* D. R. Bates, ed. New York: Basic Books, 1957, pp. 112–129.
3. E. T. EADY, "The General Circulation of the Atmosphere and the Oceans," in *The Earth and Its Atmosphere,* D. R. Bates, ed. New York: Basic Books, 1957, pp. 130–151.
4. B. J. MASON, "Meteorology," in *The Earth and Its Atmosphere,* D. R. Bates, ed. New York: Basic Books, 1957, pp. 174–203.
5. J. A. RATCLIFFE, "The Ionosphere," in *The Earth and Its Atmosphere,* D. R. Bates, ed. New York: Basic Books, 1957, pp. 204–222.
6. V. C. A. Ferrarro, "Aurorae and Magnetic Storms," in *The Earth and Its Atmosphere,* D. R. Bates, ed. New York: Basic Books, 1957, pp. 231–255.
7. S. PETTERSSEN, *Introduction to Meteorology,* 2nd ed. New York: McGraw-Hill, 1958.
8. T. A. BLAIR and R. C. FITE, *Weather Elements,* 5th ed. Englewood Cliffs, N.J.: Prentice-Hall, 1965.
9. H. S. W. MASSEY and R. L. F. BOYD, *The Upper Atmosphere.* London: Hutchinson, 1958.
10. H. R. BYERS, *General Meteorology.* New York: McGraw-Hill, 1959.
11. M. A. ELLISON, *The Sun and Its Influence.* New York: Macmillan, 1956.
12. K. KIEPENHEUER, *The Sun.* Ann Arbor, Mich.: University of Michigan Press, 1959.
13. J. A. DAY, *The Science of Weather.* Reading, Mass.: Addison-Wesley, 1966.
14. H. RIEHL, *Introduction to the Atmosphere.* New York: McGraw-Hill, 1965.
15. P. D. THOMPSON, R. O'BRIEN, and THE EDITORS OF LIFE, *Weather.* New York: Life Science Library, 1965.
16. L. J. BATTAN, *Cloud Physics and Cloud Seeding.* New York: Doubleday, 1962.
17. L. J. BATTAN, *The Nature of Violent Storms.* New York: Doubleday, 1961.
18. L. J. BATTAN, *The Unclean Sky.* New York: Doubleday, 1966.

Laboratory Exercises

The laboratory exercises presented below are designed to illustrate the geophysical principles described in the text, and are not intended to teach field or laboratory techniques in geophysics.

The exercises have been designed to fit into a biweekly laboratory program for one semester, and are coordinated with the textual material.

I. The sundial

Object: The construction and use of a sundial, and familiarization with time.

Materials: 8″ × 12″ cardboard, scissors, glue, ruler, protractor, transparent tape, magnetic compass.

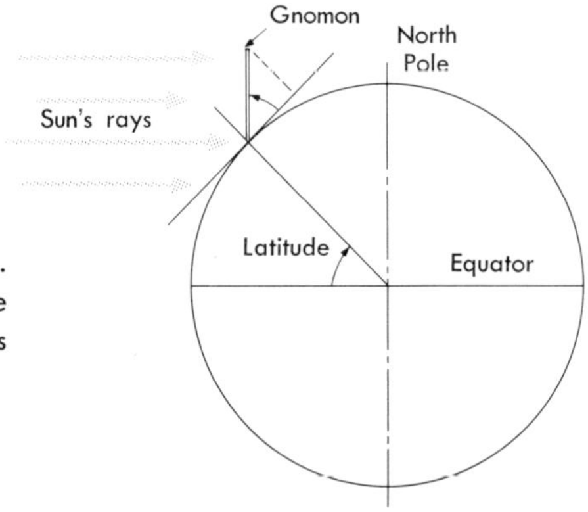

Fig. 69. Principle of the sundial. Note that the angle between the gnomon and the horizontal plane is equal to the local latitude.

Principles: The sundial is probably the most ancient of instruments used for the measurement of time. The basis for our familiar system of time, solar time, is the apparent motion of the sun, a consequence of the rotation of the earth. The sundial measures solar time by measuring the orientation of the shadow cast on the horizontal plane by a staff arranged parallel to the axis of rotation of the earth. The staff of the sundial is called the *gnomon*. As shown in Fig. 69, the gnomon of the sundial is made parallel to the axis of the earth by tilting it relative to the horizontal plane at an angle equal to the latitude of the place where it is to be used. The gnomon is then placed in the plane of the meridian, i.e., it is pointed toward the north. The shadow of the gnomon will now rotate in the horizontal plane during the day following the apparent motion of the sun.

The hour lines on the face (base) of the sundial can be calculated from a spherical trigonometric formula:

$$\tan a = (\sin l)(\tan 15h). \tag{1}$$

The formula (which we will not attempt to derive) gives the orientation of the shadow of the gnomon as a function of the latitude, l, and the number of hours, h, before or after noon. The orientation of the hour lines is represented by the angle, a, between the noon hour (the meridian and central line of the sundial) and the hour lines. The hour lines are symmetrical about the noon line. As an illustration, the angles between the hour lines and the noon line (the meridian) of the sundial at the latitude of New York, 40 degrees 46 minutes north, are listed in Table A. The time of day is expressed in 24-hour clock time. (See Fig. 70 for an illustration of the angles involved in the trigonometry of the sundial.) For any other latitude, the sundial angles may be computed from formula (1).

The linear dimensions of the sundial can be calculated from the angles in Table A. These are shown in Fig. 71, for the latitude of New York City, for a four-inch sundial.

The sundial gives *apparent solar time*. The length of the *apparent solar day* varies slightly during the year. The *mean* (average) *solar day,* which is exactly 24 hours long, is the basis for clock time. The difference, *mean solar time* minus apparent solar time, is called the *equation of time*. To

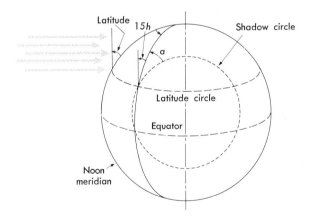

Fig. 70. Angles involved in the trigonometry of the sundial. The *hour angle,* $15h$, is the angle between the meridian of the sun (the noon meridian) and the local meridian. The plane parallel to the plane of the noon meridian, which intersects the earth's surface along the shadow circle, is the shadow plane of the gnomon. The angle between the shadow circle and the local meridian is a, the orientation angle of the hour line. The angle a of the hour line on the sun dial approaches the sun's hour angle $15h$ as the latitude increases and the shadow circle approaches the great circle of the noon meridian. Conversely, the angle a approaches zero as the latitude decreases toward the Equator. This is in accord with the trigonometric formula given in Lab. Ex. I and the data in Table A.

Table A
Sundial Angles for Latitude 40°46′ N

h	Time	a
0	1200	0
1	1100, 1300	10°25′
2	1000, 1400	21°36′
3	0900, 1500	33° 9′
4	0800, 1600	48°30′
5	0700, 1700	67°40′
6	0600, 1800	90°

determine mean solar time, *add* the equation of time (see Table B) to the apparent solar time determined from the sundial. (Note that the equation of time is sometimes positive and sometimes negative.)

Clock time is standard time, which is the mean solar time on the central meridian of the local time zone. The sun moves westward 15 degrees of longitude each hour, or one minute of longitude every four seconds. To convert mean solar time to standard (clock) time, *subtract* four seconds from mean solar time for every minute of longitude between your locality and the central meridian if you are *east* of the meridian, and *add* four seconds for every minute of longitude between the two meridians if you are *west* of the central meridian.

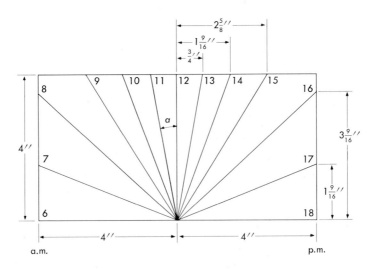

Fig. 71. Dimensions of the base of the sundial for latitude 40°46′. The radiating hour lines are labeled in 24-hour clock time.

THE SUNDIAL

Table B
Equation of Time (Approximate)

Date	Equation of time	Date	Equation of time	Date	Equation of time
Jan. 1	+ 3 min	May 1	−3 min	Sept. 1	0 min
15	+10	15	−4	15	− 5
Feb. 1	+14	June 1	−2	Oct. 1	−10
15	+15	15	0	15	−14
Mar. 1	+13	July 1	+3	Nov. 1	−16
15	+ 9	15	+5	15	−15
Apr. 1	+ 4	Aug. 1	+6	Dec. 1	−11
15	0	15	+4	15	− 4

Procedure:

1. From formula (1), compute the sundial angles for your latitude. Using this information, construct a sundial out of cardboard.

2. Set the sundial on a flat surface outdoors (weather permitting). Orient the gnomon to point to true north. (True north may be determined with a magnetic compass from the magnetic declination at your locality.) The shadow of the gnomon now gives the (approximate) *apparent solar time.*

3. Determine the *mean solar time* from the (approximate) equation of time, which is defined as the difference between mean solar and apparent solar time, i.e.,

Equation of time = mean solar time − apparent solar time.

Some values of this quantity are listed in Table B.

4. Express the time of your observation in standard time and in Greenwich Civil Time (GCT). Compare with your clock time.

5. Repeat the experiment at least twice, once in the early morning and once in the late afternoon, outside the laboratory before handing in the report.

Supplementary Exercises

1. Construct the equation-of-time curve.

2. Compute the time of local apparent noon (solar transit of the meridian) in EST and GCT on March 15, October 1, November 15, and February 15 at (a) Boston (longitude 71°10′W), (b) Jacksonville (81°40′W), (c) Caribou, Maine (68°15′W), (d) New York City (73°46′W).

3. Compute the time of local apparent noon in Central Standard Time (CST) and GCT on January 15, April 1, May 15, and December 1 at (a) Chicago (longitude 87°30′W), (b) Dallas (97°W), (c) Duluth (92°10′W).

II. Gravity

Object: The determination of gravity from the observation of a pendulum.

Materials: Meter stick, pendulum, stop watch.

Principle: A simple pendulum consists of a small mass suspended from an inextensible, weightless string. The period of a simple pendulum, for small oscillations, is given by the formula

$$T = 2\pi \sqrt{\frac{L}{g}}, \qquad (2)$$

where T is the period, L is the length of the string, g is the acceleration due to gravity, and $\pi = 3.141\ldots$ is the ratio of the circumference to the diameter of a circle. From Eq. (2) the formula for the acceleration due to gravity is found to be

$$g = \frac{4\pi^2 L}{T^2}. \qquad (3)$$

Procedure:

1. Start the mass in motion along a small arc. Measure with a stop watch the total time required for the pendulum to swing through 10 (or more) complete cycles. The period T is the total time divided by the number of cycles.

2. Measure the length of the pendulum in centimeters (cm). Compute g in centimeters per second per second (cm·sec^{-2}) from equation (3). Repeat the exercise for five different pendulum lengths.

3. Repeat the exercise five times for each pendulum length, and compute the average value of g from all the observations. Plot a graph of T^2 against L and show that the points lie on a sloping straight line.

4. Compare your results with those of the other members of the class.

III. Geomagnetism

Object: To study the magnetic field of the earth with the aid of an earth model.

Materials: Terrela (magnetic globe), dip needle, compass, protractor.

Principles: A magnetic field may be represented by lines of force, as illustrated in Fig. 72 for a simple bar magnet. A compass needle placed in the magnetic field will orient itself parallel to the magnetic lines of

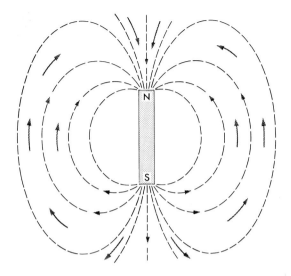

Fig. 72. Magnetic lines of force associated with a bar magnet. Arrows represent north-seeking pole of compass needle.

force, with the north-seeking pole of the compass needle pointing toward the north pole of the magnet, as shown by the arrows.

The magnetic field of the earth may be represented approximately by the magnetic field of a bar magnet placed near the center of the earth and tilted 11 degrees away from the earth's axis of rotation. This is the principle of construction of the magnetic globe. The north and south magnetic poles of the earth are the points on the surface of the globe directly in line with the axis of the bar magnet. The magnetic poles do not coincide with the geographical poles.

A compass needle free to rotate about a vertical axis on the earth (in the horizontal plane) will tend to orient itself parallel to the magnetic lines of force. If the magnetic poles corresponded to the geographical poles, the compass needle would align itself parallel to the meridians. In fact, however, the direction of the compass needle deviates from the geographic meridian by an angle called the magnetic *declination*. If the compass needle points to the east of the meridian, the declination is called degrees east, and if it points to the west of the meridian, the declination is called degrees west. Lines on the earth along which the declination is the same are called *isogons*. (Magnetic declination is sometimes called *magnetic variation*.)

A magnetic needle free to rotate about a horizontal axis (in the vertical plane) is called a dip needle. Like the compass needle, the dip needle tends to orient itself parallel to the magnetic lines of force. If the dip needle is held parallel to the magnetic meridian (i.e., parallel to the compass needle), it will show the inclination or tilt of the magnetic lines of force relative to the horizontal plane. Over the magnetic poles, the dip needle will show an inclination of 90 degrees, i.e., it will point straight down. Along the magnetic equator, the dip needle will be horizontal, with an inclination of zero.

Procedure:

1. Using a compass, determine the orientation of the compass needle (magnetic north-south line) over the entire magnetic globe relative to the geographic meridians.

2. On a polar diagram representing a polar projection of the Northern Hemisphere, plot arrows showing the orientation of the compass needle relative to the geographic meridians (Fig. 73). Draw curves parallel to the arrows to show the orientation of the magnetic field at the earth's surface.

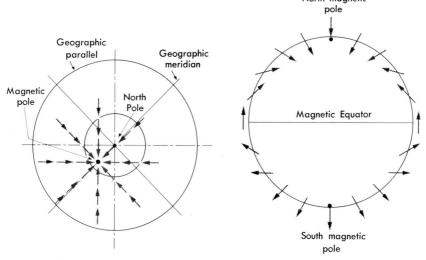

Fig. 73. Direction of compass needle displayed on polar map projection.

Fig. 74. Magnetic dip along meridian.

3. Determine the magnetic declination at New York, San Francisco, London, and Moscow.

4. Using the dip needle, determine the inclination from pole to pole along the meridian containing the magnetic poles.

5. Construct a circle representing the meridian, plot arrows showing the orientation of the dip needle, and draw curves parallel to the arrows to show the orientation of the magnetic field in a vertical plane (Fig. 74).

6. Determine the magnetic inclination at New York, San Francisco, London, and Moscow.

IV. Ocean currents

Object: To study the distribution of ocean currents in the North Atlantic Ocean.

Materials: Base map of the North Atlantic Ocean, table of current data (Table C), pen (fine point), protractor, pencil.

Principles: Measurements of ocean surface currents have been made for many years (by both scientific and commercial vessels) by observing the drift of objects on the sea surface and by evaluating navigational records. Average ocean currents are calculated by averaging these current data over a long period of time. A map of the resultant (average) currents can be constructed by plotting each current vector on a map at the proper position, as an arrow (to show direction) with barbs (to indicate current speed). Streamlines can then be drawn parallel to the current vectors with *isotachs* (lines of equal speed) to show the speed of the current.

In Table C, resultant currents for the month of July in the North Atlantic Ocean are tabulated for each five degrees of latitude and longitude.

The direction of the current is given in degrees measured clockwise from north, and shows the direction *toward* which the current is moving. [Example: a 90-degree current moves *toward* the east, a 270-degree current is *toward* the west, a 180-degree current is *toward* the south.] The speed of the current is given in *nautical miles per day*.

Procedure:

1. Plot the current data tabulated in Table C on the base map. Use an arrow to indicate current direction, and barbs to indicate current speed. A half-barb denotes a speed of five nautical miles per day and a full barb, 10 nautical miles per day. [Example: a current of direction 180° and speed 15 nautical miles per day is graphically represented as an arrow with one-and-a-half barbs, pointing toward the south.]

2. Draw streamlines parallel to the current vectors.

3. Identify the Gulf Stream, the Labrador Current, the North Atlantic Current, the Canary Current, and the Sargasso Sea.

V. Water waves

Object: To study waves in a tank.

Materials: Rectangular glass tank, water, stop watch, ruler, grease pencil, cork floats, rectangular seiche paddles, T-shaped wave generators.

Principles: The waves on the surface of the sea and on other bodies of water have probably been studied more extensively than any other hydrologic phenomena. A beautiful body of theory regarding these surface-water waves has been developed in hydrodynamics. From this theory it

Table C
Average Surface Currents in the North Atlantic Ocean: July

Lat., °N	Long., °W	Dir., deg.	Speed, naut. mi/day	Lat., °N	Long., °W	Dir., deg.	Speed, naut. mi/day	Lat., °N	Long., °W	Dir., deg.	Speed, naut. mi/day
55	55	160	11		65	20	2		50	310	6
	50	180	15		60	70	7		45	280	7
	45	180	5		55	180	1		40	250	6
	40	10	4		50	230	4		35	200	6
	35	30	2		45	180	2		30	230	3
	30	50	3		40	160	3		25	250	6
	25	30	4		35	170	5		20	230	5
	20	50	3		30	140	3				
	15	90	5		25	130	3	15	80	320	22
	10	90	6		20	180	5		75	280	17
					15	180	4		70	300	9
50	50	180	9		10	180	6		65	300	12
	45	150	7						60	310	12
	40	150	5	30	80	360	50		55	310	10
	35	40	2		75	360	2		50	280	7
	30	60	5		70	360	1		45	270	7
	25	50	3		65	10	2		40	270	15
	20	80	3		60	360	3		35	250	10
	15	90	2		55	290	4		30	270	15
	10	100	1		50	360	2		25	210	4
					45	230	2		20	180	6
45	60	180	2		40	240	8				
	55	210	5		35	240	8	10	55	320	13
	50	160	4		30	240	10		50	300	12
	45	60	10		25	180	8		45	310	8
	40	80	2		20	220	2		40	270	13
	35	100	3		15	220	6		35	270	12
	30	100	2						30	230	9
	25	80	3	25	80	40	60		25	180	3
	20	110	4		75	320	6		20	90	10
	15	130	4		70	310	4		15	120	12
	10	180	3		65	310	3				
					60	300	4	5	50	320	15
40	70	50	1		55	310	5		45	90	12
	65	70	8		50	310	4		40	30	16
	60	70	11		45	280	4		35	70	20
	55	80	13		40	260	4		30	70	13
	50	90	11		35	260	10		25	90	10
	45	70	5		30	240	3		20	30	20
	40	120	5		25	220	7		15	90	10
	35	130	4		20	230	3				
	30	120	3					0	45	280	50
	25	130	4						40	300	14
	20	130	5	20	80	40	2		35	270	21
	15	140	4		75	270	5		30	270	20
	10	180	3		70	310	10		25	270	20
					65	310	6		20	270	20
35	75	40	25		60	320	6		15	240	22
	70	40	3		55	320	6		10	290	24

is possible to predict the speed and behavior of ocean waves. We will not attempt to derive any theoretical results, but we will try to learn something about the behavior of water waves by observing them in a wave tank.

We distinguish between two types of waves: shallow-water waves and deep-water waves. In a shallow-water wave, the depth of the water is relatively small compared with the length of the wave, and shallow-water waves are therefore called long waves. In a deep-water wave, the water is relatively deep compared with the wavelength, so that these are called short waves.

Long and short waves obey different laws concerning the speed of the wave. The speed of a long wave depends only on the depth of the water. The speed of a short wave depends only on the wavelength. All long waves move through a given body of water with the same speed. All short waves move through a given body of water with different speeds, depending on the length of the waves.

Both long and short waves are called *gravity waves*. They exist and move because of the force of gravity.

The formula for the speed of a long water wave is

$$c^2 = gH, \qquad (4)$$

where c is the speed of the wave, g is the acceleration due to gravity, and H is the depth of the water.

The formula for the speed of a short water wave is

$$c^2 = \frac{gL}{2\pi}, \qquad (5)$$

where L is the length of the wave.

Waves are generated by the wind blowing across the sea surface. (An exception is the *tsunami* or *earthquake wave,* also called *tidal wave.*) The wind-generated wave usually moves, and is therefore called a *progressive wave.* Some waves, however, do not move, and are called *standing waves.* These occur mainly in enclosed water bodies, such as lakes (and bathtubs), and are called *seiches.* Although the waveform does not advance in a seiche, the water surface does move up and down.

In a standing wave, the water surface remains stationary along certain lines, while the rest of the surface oscillates. These lines are called *nodes.* Seiches with one node (uninodal seiches) and two nodes (binodal seiches) are illustrated in Fig. 75.

The period, T, of a seiche with n nodes can be calculated from the formula

$$T^2 = \frac{4M^2}{n^2 gH}, \qquad (6)$$

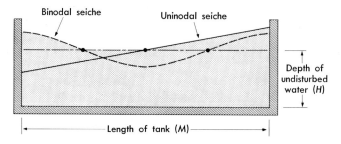

Fig. 75. Uninodal and binodal seiches in a tank.

where M is the length of the basin, g is the acceleration due to gravity, and H is the depth of the undisturbed water.

Seiches are set up by strong winds, and have been responsible on many occasions for considerable destruction of life and property by drowning and inundation.

The speed of a wave is the speed with which any phase of the waveform (the crest or trough of the wave, for example) moves horizontally, and should not be confused with the motion of the water itself. The latter may be studied by watching the motion of floating objects in the water. In Fig. 76, the movement of the water is illustrated by the single arrows, which show the motion of the water at any instant. The dashed circles show the path of an individual water particle in time as the wave moves by.

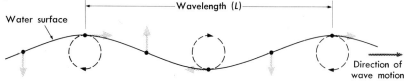

Fig. 76. Wave and water movements in a progressive wave.

The *wavelength* of a wave is the distance from crest to crest or from trough to trough, as shown in Fig. 76. The *period* of a wave is the time interval between the passages of successive crests or successive troughs. The speed of the wave is equal to the wavelength divided by the period.

Procedure: We shall study:
(A) the speed of the progressive shallow-water waves,
(B) the motion of the water as contrasted with the motion of the wave,
(C) the period of seiches.

(A) 1. Fill the tank with water to a depth of 0.5 centimeter. Place a small cork float in the center of the tank. Allow the water to come to rest.

2. Poise the T-shaped wave generator over the water at one end of the tank. Plunge it quickly into the water, and observe the solitary wave thus generated as it travels across the tank. Pull the wave generator quickly out of the water, and again observe the motion of the solitary wave.

3. Now measure with a stopwatch the time required for the wave to travel the full length of the tank. Also measure, if possible, the time required for the *reflected* wave to travel across the tank and back again.

4. Repeat the measurement on 10 waves.

5. Measure the length of the tank and the depth of the water, in centimeters.

6. Divide the length of the tank by the *average* travel time of the waves from one end of the tank to the other. The result obtained is the speed of the wave.

7. Compute the speed of the wave from the shallow-water formula, $c^2 = gH$. Assume $g = 980$ cm·sec^{-2}. Compare the observed and theoretical wave speeds.

8. Increase the depth of the water in the tank to two centimeters and repeat the experiment.

(B) 1. While carrying out procedure (A), observe the motion of the cork float.

2. Describe, illustrate, and discuss the motion of the float (water) as compared with the motion of the wave.

(C) 1. Fill the tank to a depth of 10 centimeters.

2. With grease pencil, draw a horizontal line on the glass tank coincident with the equilibrium water surface, and a vertical line bisecting the tank.

3. With the seiche paddle held vertically in the water in the middle of the tank, slowly start the water oscillating in a uninodal seiche. Move the paddle in rhythm with the seiche to increase its amplitude by resonance.

4. Quickly and carefully remove the paddle and count the time required for the seiche to execute 10 *complete* cycles.

5. Determine the average period of the seiche.

6. Compare the observed period with the theoretical period computed from the formula $T^2 = 4M^2/n^2gH$.

7. Repeat the experiment for the binodal seiche.

(a) To generate the binodal seiche, hold two seiche paddles vertically one quarter of the distance from both ends of the tank. These are the places (see Fig. 74) where the nodes will appear.

(b) Move the paddles toward each other and then away from each other in a fast, regular rhythm to generate the seiche. The water will rise and fall in the center of the tank, and, with opposite phase, at the two ends of the tank.

150   LABORATORY EXERCISES

VI. Weather observing

Object: Familiarization with meteorological instruments and observations, and their use in weather analysis.

Materials: Mercurial barometer, aneroid barograph, sling psychrometer, anemometer, wind vane, rain gage, the U.S. Weather Bureau's publication, *Explanation of the Weather Map,* cloud charts, pressure correction and reduction tables, psychrometric tables.

Principles: The purpose of this exercise is to familiarize the student with the meteorological instruments used in weather stations throughout the United States and elsewhere for routine meteorological observations. Each student will carry out, under the supervision of the instructor, a complete surface-weather observation, including all the elements normally observed by the professional weather observer. The students will then plot the weather observations, using the standard station model as described in *Explanation of the Weather Map.* (Once each week, on Sunday, the U.S. Weather Bureau prints the *Explanation of the Weather Map* on the back of the *Daily Weather Map* of the United States, published in Washington, D.C. A subscription to the *Daily Weather Map* may be purchased from the Superintendent of Documents, Government Printing Office, Washington 25, D.C.)

Procedure:

1. Determine the sea-level pressure from a reading of the mercurial barometer.

(a) Atmospheric pressure is equal to the weight of the atmosphere per unit area. The mercury barometer weighs the atmosphere against the weight of a column of mercury.

(b) A column of mercury of height h and density d exerts a downward pressure equal to $hdg,$ where g is the acceleration due to gravity. Thus the pressure of the atmosphere, which supports the column of mercury, is equal to $hdg.$

(c) To calculate the pressure, we first measure the height, $h,$ of the mercury column above the surface of mercury in a reservoir. The level of the mercury in the reservoir is adjusted, by means of a screw, to coincide with an ivory point which corresponds to zero on the height scale.

(d) The density of the mercury varies with temperature. Therefore a correction must be made for the temperature of the barometer. (Tables are available for this purpose.)

(e) Pressure decreases with increasing altitude. After the station pressure has been determined, it is reduced to sea level by adding the pressure of a fictitious column of air between the barometer and sea level. The sea-level reduction depends on the temperature of the outside air. (Tables are available for the sea-level reduction.)

WEATHER OBSERVING 151

(f) Read the height of the mercury in inches. Use tables to correct and reduce the pressure, and express the sea-level pressure in millibars.

2. From the aneroid barograph, determine the pressure tendency and the pressure characteristic, i.e., the amount and character of the pressure change during the past three hours, expressed in millibars.

3. Use the wet-bulb psychrometer outdoors to obtain the temperature, in degrees fahrenheit (from the dry-bulb thermometer), and the dew point, in degrees fahrenheit. Whirl the psychrometer, after moistening the muslin around the wet-bulb, until the wet-bulb temperature reaches a minimum value. Read both the dry- and wet-bulb temperatures, in the shade. Determine the dew point from psychrometric tables. (Psychrometric tables may be obtained from the U.S. Weather Bureau.)

4. Observe the weather, the cloud forms, amounts, and direction of motion, and the horizontal visibility. (Tables given in the *Explanation of the Weather Map* describe the various cloud and weather forms. Illustrations of the various cloud forms are shown in a photographic *Cloud Code Chart*, which can be obtained from the U.S. Weather Bureau.)

5. Determine the wind velocity (speed and direction) from the wind recorder, if one is available. Note also, and describe, the wind from your visual observation (trees bending, flags extended, smoke rising vertically, shingles flying, etc.), giving your estimate of the wind direction and speed.

6. Write out your observations in words and numbers.

7. Plot your observation as it would appear on a weather map. (See "Station Model" in *Explanation of the Weather Map*.)

INDEX

adiabatic processes, 118 f.
aerosols, 101
air, 99 ff.; *see also* atmosphere
  chemical composition, 99 ff.
  density, 105, 106
  pollution of, 101
  pressure, 105 f., 112 ff.; *see also* atmospheric pressure
  vertical motion, 119 f.
albedo, 82
anemometer, 116
angular momentum, 10
angular velocity, 10
Antarctic, 63
anticyclones, 110, 116
aphelion, 9, 10
apogee, 13
apparent solar time, 16 f., 139; *see also* solar day
asteroids, 9
astronomical unit (A.U.), 8 f.
atmosphere, 4, 99 ff.; *see also* air
  altitude and pressure, 105 ff., 108, 112 ff.
  circulation, 114
  lapse rate, 103
  structure, 99, 102, 105
  temperature, 102 f., 105, 112
  tides in, 135
  upper, 102, 105, 131 ff., 134 f.
  vertical structure, 102
atom, 1 ff.
aurora borealis, 131 ff., 135

barometer(s), 105 ff.
bathythermograph, 80
blackbody radiation, 124
Buys-Ballot's law, 110 ff.

calendar, 15, 20, 21
  month, 11, 14
catastrophism, 42 f.
centrifugal force, 9 f., 29, 32
centripetal acceleration, 28 f.
cgs system, 106
climate, 123 ff.
  continental, 129
  maritime, 129
  and solar radiation, 123
clouds, 117 ff.
cloud seeding, 118
compass, magnetic, 61 ff.

condensation level, 118
continental drift, 48
continental fracture system, 48 ff.
continental shelf, 51
continental slope, 51
continents, theories of origin, 48 f.
  and temperature distribution, 128
convection, 120
Copernican model, 6 f.
core; *see* earth
Coriolis force, 91 f., 109 f., 115 f.
cosmic rays, intensity, 134, 135
crust; *see* earth
cyclones, 110 f., 116, 119 ff., 130

day, 10, 15, 16, 18, 126
decay, radioactive, 1
declination, of moon, 96
  magnetic, 143, 144
  of sun, 16, 97
density, 8, 24, 34, 55
dew point, 102, 118 f.
diastrophism, 47 f.
dinosaurs, 45 f.
dip needle, 63, 142 f.
dipole; *see* geomagnetism
doldrums, 112
dynamo theory, 61

earth, age of, 1, 2
  axis, 20, 124
  boundary crust-mantle (Moho), 49, 54
  core, 57
  crust, 49 ff., 57
  density, 34
  dimensions, 21 ff.
  geological history, 42 ff.; *see also* geology
  interior, 55; *see also* seismic waves
  magnetic field, 62 ff., 136, 142 ff.
  mantle, 49, 57 f.
  origin, 1 ff.
  theories, 3, 4
  rotation, 15 ff.
  shape, 21 f.
  surface area, 24
  temperature, 58
earthquakes, 58 ff.
eccentricity, of planetary orbits, 7
echo sounding, 51
eclipse(s), 14

INDEX

ecliptic, 10, 16 f., 20, 126
Ekman spiral, 89
elastic rebound, 61
electromagnetism, 61 ff.
epicenter, 60
epicycles, 6
equation of time, 18, 139 ff.
Equator, geomagnetic, 64
    radius, 24
    thermal, 128
equatorial low, 112
equinox(es), 21, 121, 126
erosion, 43, 47 f.
evaporation, 68 ff., 81
exosphere, 105

faculae, 125
fault, 47, 61
flares, solar, 66
fossils, 44 ff.
Foucault pendulum, 15, 109
freezing, 68 f.
front (weather), 120 ff.

geocentric theory, 5, 6
geodesic (great circle), 24 f.
geodesy, 21 f.
geoid, 24, 33
geology (geological), 42 ff.
    eras, 45 ff.
    epochs, 46
    history, 42 ff.
    periods, 45, 46
    processes, 46 f.
geomagnetism (geomagnetic), 42, 61, 142 ff.; *see also* magnetism
    activity, 66
    declination, 63, 143, 144
    dynamo theory, 61
    equator, 64
    inclination, 64, 144
geostrophic ocean currents, 91
geostrophic wind, 110 f.
glaciers, 53, 70
gnomon, 138 f.
gravimeter, 33
gravitation, gravity, 25 f., 28 ff., 32 f., 142
    acceleration due to, 29 ff., 33
    attraction of, 10, 11
    universal law of, 7, 28 ff.
gravity waves, 147
great circle (geodesic), 24 f.
Greenwich Civil Time (GCT), 18 f., 25
gyres, 88, 116

half-life, 2 f.
heat, latent, 68

heliocentric theory, 6 f.
horse latitudes, 113
hour angle, 139
humidity, 101 f.
hurricanes, 123
hydrologic cycle, 68 ff.
hydrosphere, 68 ff.
hydrostatic equation, 111

ice, 68
ice ages, 53, 71
ice pack, Arctic, 75
igneous rocks, 46 ff.
inclination; *see* geomagnetism
index fossils, 44
insolation (incoming solar radiation), 82, 103; *see also* solar radiation, sun
    annual range, 127 f.
    and sea, 82 f.
    variations in, 125 ff.
international date line, 18
inversion, 103
ionosphere, 104 ff., 131
    layers, 104, 132, 135
    magnetic variations, 65 f.
    and radio communication, 135 f.
    Van Allen radiation belts, 105, 132 ff.
isobars, 108, 110
isobaths, 77
isogons, 143
isostasy (isostatic equilibrium), 30 f., 34 f.
isotachs, 146
isotherms, 128
isotopes, radioactive, 1 ff.
    ratio, 2

jet streams, 112

Kepler's laws, 7, 10 f.

lapse rate, 103
latitude (parallels), 24 ff., 33
level of no motion, 92; *see also* ocean currents
lithosphere, 42 ff.
longitude (meridians), 24

magma, 46
magnetic field of earth, 61 ff., 136, 142 ff.
magnetism (magnetic), 61 ff.; *see also* geomagnetism
magnetosphere, 133
mantle; *see* earth
maps, 35 ff.

INDEX 155

mass, 31 f.
  of earth, 31
mass number, 2
mean radius, 24
mean solar day, 17, 139 f.
mean solar time, 18, 141
meridian, 16
  Greenwich, 18 f.
mesopause, 104, 134
mesosphere, 104, 131
metamorphism, 48
meteorology, 99
meteors, 134
microseisms, 59
Moho, 50, 54
Mohole, 50
monsoon effect, 112, 115
moraine, glacial, 48, 53
moon, 4 f., 11 ff., 95 ff.
M-regions, 66
mountain building, 48

nautical mile, 27, 87
neap tides, 97
Newton's laws, 7, 10, 22, 28 ff.
noctilucent clouds, 134

obliquity, 16, 20
oceans, 70 ff.
  basins, 71
  bottom, 50 f., 76 f.; *see also* sediments
  and climate, 70 f.
  currents; *see* ocean currents
  depth, 71, 73 f., 74 f.
  gases in, 81 f.
  life in, 81 f., 92
  pressure, 75 f.
  salinity, 78 ff.
  storm surge, 90
  temperature, 79 f., 82
  tides, 84, 95 ff.
  waves, 92 ff., 146 ff.; *see also* seiches, water particle
ocean currents, 80 f., 84 ff., 146
  and climate, 89, 128
  deep-water, 92 f.
  and density of water, 90 ff.
  and differences in water pressure, 90 ff.
  gyres, 88 f.
  measurements of, 85 ff.
  system, 88
  variations with depth, 89
  and winds, 89 f.
orogenetic belts, 48 f.
oxygen, in sea, 81
ozone, 100

paleomagnetism, 65; *see also* magnetic field of earth
paleontology, 43
parallels; *see* latitude
pendulum, principle of, 33 f.
perigee, 13
perihelion, 9, 16
phases, moon; *see* moon
phases, water, 68 f.
photosphere, 124 f.
pibals, 117
planetary year, 9 f.
planets, 4 f., 7 ff.
plasma clouds, 133
polar front, 130
precession, 20 f.
precipitation, 69 f., 81, 117 f.; *see also* rainfall
pressure-gradient force, 91, 109 f.
psychrometer, 102
P-waves, 55 ff., 58 f., 60

radiation, 1 ff., 65 f.; *see also* atom, sun
radioactive carbon, 100
radioactive dating, 2, 100, 135
radioactive gases, 100
radioactivity, 1, 44, 101
rainfall, 69 f., 130; *see also* precipitation
rawins, 117
refraction, 55 f.
relative humidity, 117
rhumb line (loxodromic curve), 37, 40
rock stratigraphy, 43
rocketsonde, 102

salinity of ocean water; *see* ocean, salinity
sea ice, 83 f.
sea level, 70 f.
seasons, 20, 126 f.
seawater, properties of, 78 ff., 90 ff.; *sea also* ocean, salinity
sedimentation, 46 f.
sediments, on ocean floor, 50 f., 76
seiches, 94, 147 f.
seismic waves, 54 ff., 58 f.
seismograph, 54, 58
seismology, 42, 54 ff.
shadow zone, 55 ff.
sidereal day, 11, 16 f.
sidereal month, 11, 13 ff.
sidereal year, 11
solar constant, 125
solar day, 16 f.
  apparent, 16, 139

solar radiation, 123 ff.; *see also*
  insolation, sun
solar system, 5 ff., 8
solstice(s), 20 f., 126 f.
spheriod, 22 f.
  international reference, 24
spring tides, 97
standard time, 18 ff.
stadium(a), 23
station error, 26
stratopause, 134
stratosphere, 102, 104
striations (striae), 53
sun, 4 f., 123 ff.
  atmosphere, 125
  diameter, 11
  latitude, 17
  mass, 11
  radiation, 123, 125; *see also*
    insolation, solar radiation
  rotation, 11
  solar flares, 136
  storms, 135
  sunspot cycle, 136
  temperature, 124
sundial, 138 ff.
sun-earth-moon system, 11 ff.
sunspots, 125, 136
surface waves (seismic), 57, 58, 60
S-waves, 56 f., 58 f., 60
synodic month, 13
synoptic maps, 121

temperature, annual range, 127 f.; *see also* atmosphere, earth, oceans
  gradients, 112, 129
thermocline, 80
thermosphere, 104
tides; *see* ocean, tides
time, 15, 18 f.
  equation of, 18, 139 ff.
transit(s), 16 f.
triangulation, 22
tropical cyclones, 123
tropical year, 21
tropopause, 104 f.
troposphere, 103 ff.

tsunamis, 95, 147
typhoons, 123

ultraviolet radiation, 65 f., 123 ff., 132 ff.
uniformitarianism, 42
upper atmosphere; *see* atmosphere
upwelling, 89

Van Allen radiation belts, 105, 132 f.
vapor, water, 99, 117
vulcanism, 46 ff.

water, 68 ff.
  distribution over earth, 71
  evaporation, 68 ff.
  precipitation, 69 f.
  pure, 77 f.
  seawater, 78
water particle, motion, 95
water vapor, 99 ff.
water waves, 93 f., 146 ff.; *see also*
  ocean, waves
  speed, 147 ff.
  wavelength, 147 f.
weather, 117 ff.
weather forecasting, 121 f., 123
weather maps, 122, 123
weather observing, 150 f.
weight, 31 ff.
wet-bulb thermometer, 102
wind, 99, 108 ff.
  and pressure, 108, 112 ff.
  Buys-Ballot's law, 110 ff.
  convergence, 110
  and friction, 110 f.
  measurements of, 116 f.
  pressure gradient, 108; *see also*
    pressure-gradient force
  types of, 112, 113 ff., 117, 122
  variation with height, 111

year, length of, 15
  sidereal, 21
  tropical, 21

zenith angle, 23, 27